APPLIED ABSTRACT ALGEBRA

ELLIS HORWOOD SERIES IN MATHEMATICS AND ITS APPLICATIONS

Series Editor: Professor G. M. BELL, Chelsea College, University of London
(and within the same series)
Statistics and Operational Research
Editor: B. W. CONOLLY, Chelsea College, University of London

Baldock, G. R. & Bridgeman, T.	**Mathematical Theory of Wave Motion**
de Barra, G.	**Measure Theory and Integration**
Beaumont, G. P.	**Introductory Applied Probability**
Burghes, D. N. & Borrie, M.	**Modelling with Differential Equations**
Burghes, D. N. & Downs, A. M.	**Modern Introduction to Classical Mechanics and Control**
Burghes, D. N. & Graham, A.	**Introduction to Control Theory, including Optimal Control**
Burghes, D. N., Huntley, I. & McDonald, J.	**Applying Mathematics**
Burghes, D. N. & Wood, A. D.	**Mathematical Models in the Social, Management and Life Sciences**
Butkovskiy, A. G.	**Green's Functions and Transfer Functions Handbook**
Butkovskiy, A. G.	**Structure of Distributed Systems**
Chorlton, F.	**Textbook of Dynamics, 2nd Edition**
Chorlton, F.	**Vector and Tensor Methods**
Conolly, B.	**Techniques in Operational Research**
	Vol. 1: Queueing Systems
	Vol. 2: Models, Search, Randomization
Dunning-Davies, J.	**Mathematical Methods for Mathematicians, Physical Scientists and Engineers**
Eason, G., Coles, C. W., Gettinby, G.	**Mathematics and Statistics for the Bio-sciences**
Exton, H.	**Handbook of Hypergeometric Integrals**
Exton, H.	**Multiple Hypergeometric Functions and Applications**
Faux, I. D. & Pratt, M. J.	**Computational Geometry for Design and Manufacture**
Goodbody, A. M.	**Cartesian Tensors**
Goult, R. J.	**Applied Linear Algebra**
Graham, A.	**Kronecker Products and Matrix Calculus: with Applications**
Graham, A.	**Matrix Theory and Applications for Engineers and Mathematicians**
Griffel, D. H.	**Applied Functional Analysis**
Hoskins, R. F.	**Generalised Functions**
Hunter, S. C.	**Mechanics of Continuous Media, 2nd (Revised) Edition**
Huntley, I. & Johnson, R. M.	**Linear and Nonlinear Differential Equations**
Jaswon, M. A. & Rose, M. A.	**Crystal Symmetry: The Theory of Colour Crystallography**
Jones, A. J.	**Game Theory**
Kemp, K. W.	**Computational Statistics**
Kosinski, W.	**Field Singularities and Wave Analysis in Continuum Mechanics**
Marichev, O. I.	**Integral Transforms of Higher Transcendental Functions**
Meek, B. L. & Fairthorne, S.	**Using Computers**
Muller-Pfeiffer, E.	**Spectral Theory of Ordinary Differential Operators**
Nonweiler, T. R. F.	**Computational Mathematics: An Introduction to Numerical Analysis**
Oliviera-Pinto, F.	**Simulation Concepts in Mathematical Modelling**
Oliviera-Pinto, F. & Conolly, B. W.	**Applicable Mathematics of Non-physical Phenomena**
Rosser, W. G. V.	**An Introduction to Statistical Physics**
Scorer, R. S.	**Environmental Aerodynamics**
Smith, D. K.	**Network Optimisation Practice: A Computational Guide**
Stoodley, K. D. C., Lewis, T. & Stainton, C. L. S.	**Applied Statistical Techniques**
Sweet, M. V.	**Algebra, Geometry and Trigonometry for Science Students**
Temperley, H. N. V. & Trevena, D. H.	**Liquids and Their Properties**
Temperley, H. N. V.	**Graph Theory and Applications**
Twizell, E. H.	**Computational Methods for Partial Differential Equations in Biomedicine**
Wheeler, R. F.	**Rethinking Mathematical Concepts**
Whitehead, J. R.	**The Design and Analysis of Sequential Clinical Trials**

APPLIED ABSTRACT ALGEBRA

KI HANG KIM
Professor of Mathematics

and

FRED W. ROUSH
Assistant Professor of Mathematics

both of Mathematics Research Group, Alabama State University
Montgomery, Alabama, USA

ELLIS HORWOOD LIMITED
Publishers · Chichester

Halsted Press: a division of
JOHN WILEY & SONS
New York · Chichester · Brisbane · Ontario

First published in 1983 by
ELLIS HORWOOD LIMITED
Market Cross House, Cooper Street, Chichester West Sussex, PO19 1EB, England

The publisher's colophon is reproduced from James Gillison's drawing of the ancient Market Cross, Chichester

Distributors
Australia, New Zealand, South-east Asia:
JAcaranda-Wiley Ltd., Jacaranda Press
JOHN WILEY & SONS INC.,
GPO Box 859, Brisbane, Queensland 40001, Australia
Canada:
JOHN WILEY & SONS CANADA LIMITED
22 Worcester Road, Rexdale, Ontario, Canada
Europe, Africa:
JOHN WILEY & SONS LIMITED
Baffins Lane, Chichester, West Sussex, England
North and South America and the rest of the world:
Halsted Press: a division of
JOHN WILEY & SONS
605 Third Avenue, New York, NY 10016, USA

British Library Cataloguing in Publication Data
Kim, Ki Hang
Applied abstract algebra.
1. Algebra
I. Title II. Roush, Fred W.
512 QA155

Library of Congress Card No. 83-217

ISBN 0-85312-563-5 (Ellis Horwood Limited, Publishers – Library Edn.)
ISBN 0-85312-612-7 (Ellis Horwood Limited, Publishers – Student Edn.)
ISBN 0-470-27441-7 (Halsted Press)

Typeset in Press Roman by Ellis Horwood Ltd.
Printed in Great Britain by Butler & Tanner, Frome, Somerset.

Table of Contents

Preface

This book is intended for a two-semester course in abstract algebra. In addition to the standard concepts of sets, groups, rings, fields, and so on we also give applications. We hope that for many students this may increase the interest of the course and help explain why abstraction and generality are useful.

Distinctive features of the text include the following:

(a) It emphasizes relational structures as well as operational structures. A unified approach is given by the parallel treatment of basic ideas valid for all types of structures.
(b) The book is comprehensive in treating important topics not usually emphasized, and the different results are interwoven in a way that reveals the interconnections between the various concepts.
(c) It is written in a down-to-earth style, with examples giving simple illustrations of most concepts.
(d) It is organized into a few chapters dealing with major areas of abstract algebra, such as groups or rings, these being divided into many sections which can be covered in 1–3 lectures.
(e) It contains exercises graded in three levels in order to accommodate students with varying mathematical backgrounds.
(f) The book contains applications to many areas which illustrate the usefulness of abstract algebra.
(g) It contains some very important open problems in algebra for which students can understand the problem after this course. This will give some indication of what research is like in mathematics.

To give a variety of applications we have put in several sections on semigroups and their applications and group representations. However, the student or reader need not go through all this material. The selection should be left up to the interest of the individual or instructor. The reader is assumed to have had calculus and some exposure to linear algebra or matrix theory.

We have put in exercises immediately after each section, divided into three levels. It is strongly recommended that each student go through all the Level 1 exercises. Level 2 exercises require some experience in proving theorems, and develop this skill. Level 3 exercises are in many cases very challenging, and include important theorems not in the text itself. They are for the student who is deeply interested in mathematics and wants to go beyond the material presented in the text itself.

The authors are happy to acknowledge an unknown official referee for numerous very constructive criticisms and suggestions. Also, both authors are grateful to Mrs Kay Roush for her diligent and accurate proofreading.

Ki Hang Kim
Mundok, Korea

Fred William Roush
Montgomery, Alabama

CHAPTER 1

Sets and binary relations

In this chapter we cover the most basic types of algebraic structures: binary relations on sets. We first review material on the theory of sets itself.

In contrast to binary operations like addition and multiplication from two things yield a third, sum or product, a binary relation only compares two quantities, as $x < y$ or $x = y$.

There are three kinds of binary relations we consider at greater length. A ***function*** $y = f(x)$ expresses the fact that x determines y, as a cause determines its effect. Functions are of importance in many branches of mathematics. By means of functions different structures can be compared.

Equivalence relations like the geometric idea of similarity express the idea that two elements are the same in some respect.

Order relations like 'greater than' or 'X is a subset of Y' establish a comparison of lesser to greater. The key property of order relations is ***transitivity***: if $x < y$ and $y < z$ then $x < z$. There are many varieties of order relations such as ***semiorders***, ***preorders*** (***quasiorders***) and ***weak orders***.

Finally we prove a well-known theorem in social choice theory using those results. The preferences of any individual of a group over a list of possible actions of the group can be represented by an order relation. A group choice method can be taken as a certain type of function. The theorem shows certain kinds of group choice methods are possible only if there are no more than two alternatives.

1.1 SETS

In application of quantitative mathematics such as arithmetic, algebra, geometry, calculus, differential and integral equations, and linear algebra, objects are represented by numbers or n-tuples of numbers, which are measurements of that object. Relationships among objects such as joint motion under gravitational or electromagnetic forces, are represented by relationships among numbers such as functions or equations or by relationships among functions.

In applications of nonquantitative mathematics objects are frequently represented by ***sets*** or ***elements of sets***. Mathematically little can be said about a nonnumerical object except its relationship to other objects. Mathematics can deal with nonnumerical relationships of a formal nature. All an object in a set has is a name.

Sets are collections of objects. The simplest way to specify a set is to list its elements as {orange, apple, peach} or {A, E, I, O, U}. A second way is to give a condition describing precisely those objects which are members of the set. An example is $\{x : x \text{ is a positive odd integer less than } 20\}$.

In a sense the theory of sets (with logic) is the foundational part of mathematics. That is, numbers can be described in terms of sets. For instance for the number 7 one can build a set with exactly 7 elements, and use it to compare other sets with, to see if they have 7 elements. This takes care of positive integers.

From positive integers one can define negative integers, fractions, and real numbers successively. Then geometry can be done from numbers using coordinates. So all of mathematics can be derived from set theory.

The basic operations with sets are ***union***, ***intersection***, and ***complement***.

DEFINITION 1.1.1. Let $\mathcal{F}$ be a family (set) of sets. Then the *union* of the members of $\mathcal{F}$, denoted

$$\bigcup_{A \in \mathcal{F}} A \quad \text{or} \quad \bigcup_{\mathcal{F}} A$$

is the set of objects which are members of at least one set of $\mathcal{F}$.

The *intersection* of the members of $\mathcal{F}$, denoted

$$\bigcap_{A \in \mathcal{F}} A \quad \text{or} \quad \bigcap_{\mathcal{F}} A$$

is the set of all objects which are members of every set of $\mathcal{F}$.

For two sets A, B the union and intersection are written $A \cup B$ and $A \cap B$.

The set A is ***contained*** in the set B if all members of A are also members of B. This is denoted $A \subset B$ (or $B \supset A$) and it can also be said that A is a ***subset*** of B. To prove a statement that one set (A) is contained in another (B), the straightforward approach is to let $x \in A$ and then prove $x \in B$.

If A and B have exactly the same members, we say $A = B$. To prove a statement that $A = B$ one can first prove $A \subset B$ and then $B \supset A$. Or one can show $A = C$ and $C = B$ for some other set C. If $A \subset B$ but $A \neq B$ then A is called a ***proper subset*** of B. Occasionally it is possible to show directly that $x \in A$ if and only if $x \in B$.

Suppose we are considering subsets of a fixed set. We call this set the *universal set* or *universe of discourse.* We denote this set by $\mathcal{U}$. If we are considering plane geometric figures, $\mathcal{U}$ could be the set of all points of the plane.

DEFINITION 1.1.2. The *complement* $\tilde{A}$ of $A \subset \mathcal{U}$ is the set of all elements of $\mathcal{U}$ not in A.

The relative complement of any two sets is also important.

DEFINITION 1.1.3. The *relative complement* $A \backslash B$ (or $A - B$, $A \sim B$) is the set of all elements in A but not in B.

The following laws hold for the operations of set theory.

(1) Commutativity

$$A \cup B = B \cup A, \qquad A \cap B = B \cap A$$

(2) Associativity

$$A \cup (B \cup C) = (A \cup B) \cup C, \qquad A \cap (B \cap C) = (A \cap B) \cap C$$

(3) Distributive Laws

$$A \cap (B \cup C) = (A \cap B) \cup (A \cap C) \quad A \cup (B \cap C) = (A \cup B) \cap (A \cup C)$$

(4) de Morgan's Law

$$\widetilde{A \cup B} = \tilde{A} \cap \tilde{B}, \qquad \widetilde{A \cap B} = \tilde{A} \cup \tilde{B}$$

(5) Idempotency

$$A \cup A = A, \qquad A \cap A = A$$

(6) Double Negative

$$\tilde{\tilde{A}} = \mathrm{A}$$

These laws generalize from operations with just two sets to unions and intersections of arbitrary families of sets.

The *null set* (*empty set*) $\emptyset$ is the set $\{\ \}$ containing no elements.

DEFINITION 1.1.4. Two sets A, B are *disjoint* if and only if their intersection is the null set.

The following laws apply to the null set and the universal set $\mathcal{U} = \tilde{\emptyset}$:

$$A \cup \emptyset = A, \qquad A \cap \emptyset = \emptyset$$

$$A \cup \mathcal{U} = \mathcal{U}, \qquad A \cap \mathcal{U} = A$$

The *cardinality* (*order*) $|A|$ of a set A is the number of elements of A, if this number is finite. Cardinality has also been defined for infinite sets.

There is no set of everything although a different concept, a class of all sets can be defined. If there were a set of all sets there would also be a set $\mathsf{S} = \{$all sets which are not members of themselves$\}$. But if $\mathsf{S} \in \mathsf{S}$ then $\mathsf{S} \notin \mathsf{S}$. If $\mathsf{S} \notin \mathsf{S}$ then $\mathsf{S} \in \mathsf{S}$. This is a contradiction, known as ***Russell's Paradox***: *both* $\mathsf{S} \in \mathsf{S}$ *and* $\mathsf{S} \notin \mathsf{S}$ *are impossible*. Therefore there is no set of all sets.

Instead sets are built from specific other sets. For example, for a set A we may define the set of all objects of A having property $\boldsymbol{P}$. This is a subset of A for any property $\boldsymbol{P}$. Sets can also be constructed by unions, intersections, taking the set of all subsets of another set, and two processes considered later: ***Cartesian product*** and the ***axiom of choice.***

EXERCISES

Level 1

For sets $\mathcal{U} = \{\text{positive integers less than } 8\}$, $A = \{1,3,5,7\}$, $B = \{3,4,6,7\}$, $C = \{4,5,6,7\}$, compute the following:

1. $A \cup B$.
2. $\widetilde{A \cap B}$.
3. $\widetilde{A \cup (B \cap C)}$.
4. $A \cap (B \cup C)$.
5. $(A \cap B) \setminus (A \cap C)$.
6. Give additional examples of sets.

Level 2

1. Verify the associative law for union for a specific example (such as that above).
2. Verify the distributive law $A \cap (B \cup C) = (A \cap B) \cup (A \cap C)$ for a specific set.
3. Which laws of sets are valid for arithmetic if one replaces $\cup$ by $+$, $\cap$ by $\times$, $\emptyset$ by 0, and $\mathcal{U}$ by 1?
4. Give examples to show the other laws of sets fail for arithmetic.
5. Prove if $A \subset B$ and $B \subset C$ then $A \subset C$.
6. Prove $\widetilde{A \cup B} = \tilde{A} \cap \tilde{B}$.

Level 3

1. Prove the distributive law $A \cup (B \cap C) = (A \cup B) \cap (A \cup C)$.
2. Deduce the other distributive law from this using de Morgan's law.
3. How many subsets does a set of n elements have?
4. When is $|A \cup B| = |A| + |B|$?
5. Give a formula for $|A \cup B|$ if $|A \cap B| \neq 0$.
6. Explain why the method of Venn diagrams provides adequate proofs of equations involving at most 3 sets. (There are 8 classes of elements x in any such 3 sets according as $x \in A$ or not, $x \in B$ or not, $x \in C$ or not).
7. Show we can test any law for 3 sets by considering the example used in Level 1 above.

1.2 BINARY RELATIONS

The ***Cartesian product*** of n sets is the set of ordered n-tuples from them, meaning things like $(1, 2, 3, 4, 7)$ with entries. It is used in a number of mathematical constructions. The Cartesian product of the real numbers with itself is the set of pairs (x, y) of coordinates in geometry, and is the basis for the name (Cartesian geometry, after René Descartes).

DEFINITION 1.2.1. The *Cartesian product* $A_1 \times A_2 \times \ldots \times A_n$ of n sets is the set of all ordered n-tuples $(x_1, x_2, \ldots, x_n)$ for which $x_i \in A_i$, $i = 1$ to n.

EXAMPLE 1.2.1. $\{a, b, c\} \times \{a, b\} = \{(a, a), (a, b), (b, a), (b, b), (c, a), (c, b)\}$.

The Cartesian product of finite sets $A_1, A_2, \ldots, A_n$ has $|A_1||A_2| \ldots |A_n|$ elements. It has other properties related to products, such as the distributive laws $(A \cup B) \times C = (A \times C) \cup (B \times C)$ and $C \times (A \cup B) = (C \times A) \cup (C \times B)$. They occur in defining binary operations like addition and multiplication, and in studying binary relations like $=$ and $<$. A message or sequence of n symbols each chosen from a set A, like the alphabet, is equivalent to an element of the set $A \times A \times \ldots \times A = A^n$.

DEFINITION 1.2.2. A *binary relation* from a set A to a set B is a subset of the Cartesian product $A \times B$.

EXAMPLE 1.2.2. The subset $S = \{(x, y) : x^2 < y^2, x, y \in \mathbf{R}\}$ is a binary relation. Here $\mathbf{R}$ denotes the set of all real numbers. It is considered a representation of the relationship $x^2 < y^2$.

For every precisely defined relationship between numbers or objects there exists a binary relation, which is the set of ordered pairs x, y such that x has the relationship in question to y. For any binary relation we conversely have the relationship that x and y are such that (x, y) belongs in the binary relation. All the logical properties of the relationship are true of the binary relation also. For instance the transitive property of inequality if $x < y$ and $y < z$ then $x < z$ is the property if $(x, y) \in \mathbf{R}$ and $(y, z) \in \mathbf{R}$ then $(x, z) \in \mathbf{R}$.

Two types of binary relations are equivalence relations and functions. An ***equivalence relation*** is a relation like $x = y$ or in geometry, $x \simeq y$ (congruence) or $x \sim y$ (similarity) which says that x, y are the same in a certain respect.

DEFINITION 1.2.3. A binary relation R on a set X is *reflexive* if and only if $(x, x) \in R$ for all $x \in X$. That it is *symmetric* means $(x, y) \in R$ if and only if $(y, x) \in R$. That it is *transitive* means if $(x, y) \in R$ and $(y, z) \in R$ then $(x, z) \in$ R.

DEFINITION 1.2.4. A reflexive, symmetric, transitive binary relation is called an *equivalence relation*.

EXAMPLE 1.2.3. The relation $x = y$ has these three properties. For all x, $x = x$. If $x = y$ then $y = x$. If $x = y$ and $y = z$ then $x = z$. It is an equivalence relation.

Equivalence relations can be expressed as follows. The set X is divided into classes. Then two members are equivalent if and only if they belong to the same class. For the relation equality the classes are single elements, so that two equivalent elements must be identical.

A division of a set into disjoint nonempty subsets is called a ***partition***.

DEFINITION 1.2.5. A family F of nonempty subsets of a set X is a *partition* if and only if

(1) $$\bigcup_{A \in F} A = X$$

(2) for $A, B \in X$ if $A \neq B$ then $A \cap B = \emptyset$

EXAMPLE 1.2.4. The set $\{1, 2, 3, 4, 5\}$ can be partitioned into the subsets $\{1, 3\}, \{2, 5\}, \{4\}$.

THEOREM 1.2.1. *For any equivalence relation R on a set X, the family of subsets $S = \{y \in X: (x, y) \in R\}$ for $x \in X$ form a partition of X. Every partition of X arises in this way from a unique equivalence relation.*

Proof. The sets S are nonempty since $(x, x) \in R$ and thus $x \in S$. This also implies $\cup S$ contains every x so $\cup S = X$. Let S, T be the equivalence classes of x, y. Suppose $S \cap T \neq \emptyset$. Let $z \in S \cap T$. Then $(x, z) \in R$ and $(y, z) \in R$. For any $w \in S$ we have $(x, w) \in R$. Since $(x, z) \in R$, $(z, x) \in R$ by symmetry. Since $(y, z) \in R$ and $(z, x) \in R$, $(y, x) \in R$ by transitivity. Since $(y, x) \in R$ and $(x, w) \in R$, $(y, w) \in R$. So $w \in T$. So if $w \in S$ then $w \in T$. Thus $S \subset T$. By symmetry $T \subset S$. So $S = T$. This proves any two classes S, T which are not disjoint are equal. Therefore $\{S\}$ is a partition.

To a partition F associate the binary relation $R = \{(x, y)$: for some $A \in F$, $x \in A$ and $y \in A\}$. This is an equivalence relation. For any $x \in X$, $x \in A$ for some A since the union of A is X. Then $x \in A$ and $x \in A$. So $(x, x) \in R$. If $x \in A$ and $y \in A$ then $y \in A$ and $x \in A$. If $(x, y) \in R$ and $(y, z) \in R$ then $x \in A$, $y \in A$, $y \in B$, $z \in B$ for some $A, B \in F$. Then $y \in A \cap B$ so $A \cap B \neq \emptyset$. So $A = B$. So $x \in A$ and $z \in B = A$. So $(x, z) \in A$.

It is straightforward to show that R does give rise to the partition F. Let $x \in A$. Then it can be shown $S = A$. Uniqueness of R can also be shown. □

EXERCISES

Level 1

1. Write out all partitions of $\{1, 2\}$. There are two: one with a single 2-element subset and one with two 1-element subsets.

2. Write out all partitions of $\{1, 2, 3\}$. There are one with a 3-element set, three with a 2-element set and a 1-element set, and one with three 1-element sets.
3. Let R be the binary relation on real numbers that x, y have the same sign or are both zero. Is this an equivalence relation? Give examples of the three properties.
4. What is the equivalence class of 1 in this relation? It will be $\{x : x, 1 \text{ have the same sign}\}$. What is this?
5. What are the equivalence classes of -1 and of 0?
6. What partition of the real numbers corresponds to this relation?
7. List 16 binary relations (not all equivalence relations) on $\{1, 2\}$. Simply list the possible sets of ordered pairs, such as $\{(1, 1), (1, 2), (2, 2)\}$.

Level 2

1. Show that if R is an equivalence relation on S and $T \subset S$ then R gives an equivalence relation on T.
2. What are the equivalence classes of T?
3. Show that on a Cartesian product $A \times B$ the relation $(a_1, b_1) R (a_2, b_2)$ if and only if $a_1 = a_2$ is an equivalence relation. What are the equivalence classes?
4. Show the relation $x^2(1-x^2) = y^2(1-y^2)$ is an equivalence relation. Generalize this.
5. What are the equivalence classes of the relation in the above exercise?
6. Show the universal relation $U = \{(x, y) \colon x \in \mathbf{R}, y \in \mathbf{R}\}$ is an equivalence relation.

Level 3

1. Let $S(n, k)$ denote the number of partitions of $\{1, 2, \ldots, n\}$, or any n-element set having exactly k distinct members. This is the same as the number of equivalence classes on an n-element set having k members. What are $S(n, n)$ and $S(n, 1)$? The $S(n, k)$ are called *Stirling's numbers of the second kind*.
2. Find a formula for $S(n, n-1)$. Describe all partitions with $n-1$ classes.
3. Find a formula for $S(n, 2)$. Describe all partitions with 2 classes.
4. Tell why $S(n+1, k) = kS(n, k) + S(n, k-1)$. (Add an element to any existing equivalence class or make it a new class.)
5. Compute $S(n, k)$ for $n \leqslant 5$ using this formula laid out in a table

   2 1 1

   3 1 3 1

   Each time to get an entry in a lower row in column k, add the entry to upper left of it and k times the entry just above it.
6. How many symmetric, reflexive binary relations are there on an n-element set? The number of transitive relations is an unsolved problem.

1.3 FUNCTIONS

A ***function*** is a relationship in which a quantity uniquely determines a second quantity. For instance x uniquely determines $x^2 + x + 1$ but it does not determine y if the relation is $x < y$.

DEFINITION 1.3.1. A *partial function* from S to T is a binary relation R such that if $(x, z) \in R$ and $(x, y) \in R$ then $y = z$. The *domain* of a partial function R is $\{x : \text{for some } y \in T, (x, y) \in R\}$. A *function* is a partial function with domain S. The set T is called the *range.*

EXAMPLE 1.3.1. On the real numbers $1/x$ is a partial function with domain $\{x : x \neq 0\}$.

A partial function differs from a function only in that it may not be defined for all x. The following are situations in which some quantity can be regarded as a function.

EXAMPLE 1.3.2. If x causes y then y is in some sense a function of x.

EXAMPLE 1.3.3. The output of a machine is a function of its input and its internal state.

EXAMPLE 1.3.4. The result of a measurement on a system is a function of the state of the system.

EXAMPLE 1.3.5. Any fixed property of a person could be considered a function of his name, since the name uniquely determines the person (ignoring people with identical names).

EXAMPLE 1.3.6. The outcome of an election is determined by the votes, essentially the preferences, of individual voters.

EXAMPLE 1.3.7. A *binary operation* on a set S, such as addition, is a function from $S \times S$ to S where $f(a, b) = a + b$.

EXAMPLE 1.3.8. A ***homomorphism*** from one multiplicative system to another is a function f such that $f(xy) = f(x)f(y)$. An example is x^n from the real numbers to itself.

EXAMPLE 1.3.9. A map of a region can be considered as a function for which any point of the region determines a corresponding point on the map.

Because of this last example, functions in general are called *maps* or *mappings.*

They may also be thought of as assignments of an element of T to each element of S.

EXAMPLE 1.3.10. The function $f(1) = 1$, $f(2) = 3$, $f(3) = 2$ can be represented as

$$\begin{array}{ccc} 1 & \longrightarrow & 1 \\ 2 & \searrow\!\!\nearrow & 2 \\ 3 & \nearrow\!\!\searrow & 3 \end{array}$$

For a function f, and element $x \in S$, $f(x)$ denotes the unique number such that $(x, f(x)) \in f$.

There is a method of obtaining a product of functions on general sets called ***composition***. The second function is applied to the result of the first function. In algebraic terms, the first function is substituted into the second.

DEFINITION 1.3.2. For binary relation R from A to B and a binary relation S from B to C, the *composition* $R \circ S$ is $\{(a, c): a \in A, c \in C$ and for some $b \in B$, both $(a, b) \in R$ and $(b, c) \in S\}$.

EXAMPLE 1.3.11. For a transitive binary relation R, $R \circ R$ is contained in R. If R is transitive and reflexive then $R \circ R = R$.

EXAMPLE 1.3.12. If R is $x^2 + 1 = y$ and S is $x < y$ then $R \circ S$ is $x^2 + 1 < y$.

For a function this definition takes this form.

DEFINITION 1.3.3. The *function* $(f \circ g)(x)$ is $g(f(x))$. The domains and ranges are as above.

EXAMPLE 1.3.13. If $f(x)$ is $x^2 + x + 2$ and $g(x)$ is $3x + 4$ then $g(f(x))$ is $3(x^2 + x + 2) + 4$.

PROPOSITION 1.3.1. *Composition of binary relations is associative.*

Proof. Let R, S, T be binary relations from A to B, B to C, C to D. Let (a, d) belong to $R \circ (S \circ T)$. Then there exist $(a, b) \in R$, $(b, d) \in S \circ T$ by definition of composition. Thus there exist $(b, c) \in S$, $(c, d) \in T$. Therefore $(a, c) \in R \circ S$. Since $(c, d) \in T$, $(a, d) \in (R \circ S) \circ T$. This proves $R \circ (S \circ T) \subset (R \circ S) \circ T$. A similar argument proves the reverse inclusion. □

DEFINITION 1.3.4. The *identity function* ι from a set to itself is the function defined by $\iota(x) = x$ for every x.

PROPOSITION 1.3.2. *For any function f, and the appropriate identity functions* $\iota_S, \iota_T, f \circ \iota_T = \iota_S \circ f = f$.

Proof. $(\iota_S \circ f)(x) = f(\iota_S(x)) = f(x)$ and $(f \circ \iota_T)(x) = \iota_T(f(x)) = f(x)$. □

DEFINITION 1.3.5. That a function f: $S \to T$ is *one-to-one* (or 1-to-1, 1-1) means if $x \neq w$ then $f(x) \neq f(w)$.

In terms of relations this is, if $(x, y) \in f$ and $(w, y) \in f$ then $x = w$.

EXAMPLE 1.3.14. If $f(2) = f(3)$ then f would not be 1-1.

DEFINITION 1.3.6. That a function f is *onto* means if $y \in T$ then there exists $x \in S$ such that $f(x) = y$.

EXAMPLE 1.3.15. There is a function from the positive integers into itself which is 1-1 but not onto, given by $f(x) = x + 1$.

A function which is both 1-1 and onto is called a ***1-1 correspondence,*** or ***isomorphism*** of sets.

EXAMPLE 1.3.16. The function $f(x) = -x$ gives a 1-1 correspondence between positive integers and negative integers.

For finite sets any two of these conditions implies the third: f is 1-1, f is onto, $|S| = |T|$.

PROPOSITION 1.3.3. *A function* f: $S \to T$ *is a 1-1 correspondence if and only if there exists a function* g: $T \to S$ *such that* $f \circ g$ *and* $g \circ f$ *are both identity functions.*

Proof. Suppose g exists. If $f(x) = f(w)$ then $g(f(x)) = g(f(w))$. But $(f \circ g)(x) = \iota(x) = x$. So $x = w$. This proves f is 1-1. For any $y \in T$, $f(g(y)) = (g \circ f)(y) = \iota(y) = y$. So f is onto.

Conversely suppose f is 1-1 and onto. By onto, for each $y \in T$ there exists $x \in S$ with $f(x) = y$. Choose such an x and define $g(y) = x$. Then $f(g(y)) = f(x) = y$. So $g \circ f$ is an identity function. Thus $f(g(f(x))) = f(x)$. Since f is 1-1, $g(f(x)) = x$. □

The function g is called the *inverse* of f. The inverse function of f is denoted f^{-1}.

We conclude this section with the ideas of n-ary operation and relation.

DEFINITION 1.3.7. An *n-ary relation* on a set S is a subset R of $S \times S \times \ldots \times S$ (n factors). An *n-ary operation* is a function f: $S \times S \times \ldots \times S$ to S. An *algebraic structure* is a finite collection of n-ary relations and operations.

EXAMPLE 1.3.17. For $n = 2$ we have binary relation and binary operations, such as addition, subtraction, multiplication and division. The relation x is between y and z on a straight line is a 3-ary (ternary) relation. The operation of finding the average of n numbers is an n-ary operation.

Abstract algebra is the study of algebraic structures in this sense.

DEFINITION 1.3.8. An *isomorphism* between two algebraic structures on sets S, T is 1-1, onto function $h: S \to T$ such that (1) for all corresponding n-ary operations f_i, g_i we have $h(f_i(s_1, s_2, \ldots, s_n)) = g_i(h(s_1), h(s_2), \ldots, h(s_n))$, (2) for all corresponding n-ary relations R_1, R_2, we have $(s_1, s_2, \ldots, s_n) \in R_1$ if and only if $(h(s_1), h(s_2), \ldots, h(s_n)) \in R_2$.

EXAMPLE 1.3.18. ***Complex conjugation*** $x + iy \to x - iy$ is an isomorphism of the complex numbers to itself. The real numbers under the binary relation $<$ and the operations $+$, $\times$ has no isomorphisms to itself except the identity.

EXERCISES

Level 1

1. Is this binary relation a function on $\{1, 2, 3\}$: $\{(1, 1), (3, 3)\}$?
2. Is this binary relation a function on $\{1, 2, 3\}$: $\{(1, 1), (1, 2), (2, 1), (3, 3)\}$?
3. Write the composition of f with itself where $f(1) = 2$, $f(2) = 3$, $f(3) = 1$.
4. Give an example of a function from $\{1, 2, 3\}$ to $\{1, 2, 3, 4\}$ which is 1-1 but not onto.
5. Can a function from $\{1, 2\}$ to $\{1, 2, 3\}$ be onto?
6. Can a function from $\{1, 2, 3, 4\}$ to $\{1, 2, 3\}$ be 1-1?

Level 2

1. Write out the functions from the set $\{1, 2, 3\}$ to the set $\{1, 2\}$.
2. Show by an example that $f \circ g \neq g \circ f$ in general.
3. What is the composition of x^n and x^m?
4. What is the inverse function to x^n?
5. Prove a composition of 1-1 functions is 1-1.
6. Prove a composition of onto functions is onto.

Level 3

1. How many functions exist from a set of n elements to a set of n elements?
2. What is the inverse function to $y = x^2 + x + 1$?
3. What is the inverse function to $y = x + 1/x$?
4. Find a 1-1 correspondence from positive integers to all integers.
5. Show this is a 1-1 correspondence from pairs of positive integers to positive integers:

$$f(n, m) = \frac{(n + m)(n + m - 1)}{2} + m$$

6. Prove the set of rational numbers under $+$, $\times$ has no isomorphisms to itself except the identity. Show first $f(0) = 0$, $f(1) = 1$, $f(n) = n$ by induction.

1.4 ORDER RELATIONS

A third important class of binary relations, in addition to equivalence relations and functions is the class of *partial orders* (and related structures). One example of a strict partial order is $x < y$ on any subset of the real numbers. Another is *set* ***inclusion***: $S \subsetneqq T$. These share the common property of transitivity. If $x < y$ and $y < z$ then $x < z$. If $X \subset Y$ and $Y \subset Z$ then $X \subset Z$. Transitivity is the fundamental property of an order relation.

In addition we have $z \nless z$ for all z and the same *is* true of $\subsetneqq$. This is called ***irreflexivity***. And if $a < b$ then $b \nless a$ (***antisymmetry***). The same holds for sets. The trichotomy property that $x < y$, $y < x$ or $y = x$ does not hold for sets.

EXAMPLE 1.4.1. $\{1\} \not\subset \{2\}$, $\{2\} \not\subset \{1\}$, $\{1\} \neq \{2\}$.

Let $x\,R\,y$ denote $(x, y) \in R$.

DEFINITION 1.4.1. A binary relation R is *irreflexive* (*reflexive*) if $x\,R\,x$ is false (true) for all x.

DEFINITION 1.4.2. That a binary relation R is *antisymmetric* means if $x \neq y$ and $x\,R\,y$ then not $y\,R\,x$.

DEFINITION 1.4.3. An irreflexive transitive binary relation is a *strict partial order.* A reflexive antisymmetric transitive binary relation is a *partial order.*

EXAMPLE 1.4.2. The relation $<$ is a strict partial order, but $\leqslant$ is not a strict partial order.

PROPOSITION 1.4.1. *Any strict partial order is antisymmetric. If P is a strict partial order then $P \cup \iota$ is a partial order. Conversely if P is a partial order $P \backslash \iota$ is a partial order. Here ι denotes an identity function* $\{(x, x) : x \in X\}$.

Proof. Let P be a strict partial order. Let $a\,P\,b$, $b\,P\,a$. Then $a\,P\,a$ by transitivity. This is a contradiction. So $P \cup \iota$ is antisymmetric. Let $(a, b) \in P \cup \iota$, $(b, c) \in P \cup \iota$. If both are in P, so is (a, c). If $a = b$ or $b = c$ then $(a, c) = (b, c)$ or (a, b) which belong to P. Thus $P \cup \iota$ is transitive. Since $\iota \subset P \cup \iota$, $P \cup \iota$ is reflexive. So $P \cup \iota$ is a partial order.

If P is a partial order, $\iota \subset P$ by reflexivity. The relation $P \backslash \iota$ is irreflexive since ι is removed. Let $(a, b) \in P \backslash \iota$, $(b, c) \in P \backslash \iota$. Then $(a, c) \in P$. Suppose $(a, c) \in \iota$. Then $(a, b) \in P$, $(b, a) \in P$, $a \neq b$. This contradicts antisymmetry. □

So a partial order has a corresponding strict partial order related to it in exactly the way $<$ is related to $\leqslant$.

DEFINITION 1.4.4. A *preorder* is a reflexive, transitive binary relation. Preorders are also known as *quasiorders*.

EXAMPLE 1.4.3. The relation that one triangle has area at least as large as another is a preorder.

Every partial order is a preorder, but some preorders are not partial orders. However, every preorder can be reduced to a partial order on certain equivalence classes.

DEFINITION 1.4.5. For a preorder Q, the *indifference relation* Q_{D} is $\{(x, y): (x, y) \in Q \text{ and } (y, x) \in Q\}$. The *strict order* Q_{S} is $\{(x, y): (x, y) \in Q \text{ but } (y, x) \notin Q\}$.

EXAMPLE 1.4.4. In the preceding example Q_{D} would be the relation that two triangles have equal areas, Q_{S} that the area of one exceeds the area of the other.

PROPOSITION 1.4.2. *The indifference relation Q_{D} of a preorder Q is an equivalence relation. There exists a unique partial order T on the equivalence classes of Q_{D} such that $(\bar{x}, \bar{y}) \in T$ if and only if $(x, y) \in Q$.*

Proof. Since Q is reflexive, $(x, x) \in Q_{\text{D}}$. So Q_{D} is reflexive. By definition, Q_{D} is symmetric. If $(a, b) \in Q_{\text{D}}$ and $(b, c) \in Q_{\text{D}}$ then they belong to Q. So $(a, c) \in Q$. Since $(b, a) \in Q$, $(c, b) \in Q$, by transitivity $(c, a) \in Q$. Thus $(a, c) \in Q_{\text{D}}$. This proves Q_{D} is an equivalence relation.

Define P by $(\bar{x}, \bar{y}) \in P$ if and only if $(a, b) \in Q$ for some representatives $a \in \bar{x}$, $b \in \bar{y}$. Then if $a_1 Q_{\text{D}} a$, $b_1 Q_{\text{D}} b$ we have $(a_1, a) \in Q$, $(a, b) \in Q$ and by transitivity $(a_1, b) \in Q$. By transitivity from $(a_1, b) \in Q$, $(b, b_1) \in Q$ we have $(a_1, b_1) \in Q$. So $(\bar{x}, \bar{y}) \in P$ if and only if $(x, y) \in Q$ for any $x \in \bar{x}$, $y \in \bar{y}$.

It remains to be shown that P is a partial order. Transitivity and reflexivity of P follow from that of Q. Antisymmetry of P follows from the fact that if $\bar{b} \, P \, \bar{a}$ and $\bar{a} \, P \, \bar{b}$ then $a \, Q_{\text{D}} \, b$ so $\bar{a} = \bar{b}$. □

Therefore preorders can be regarded as partial orders on sets of equivalence classes.

EXAMPLE 1.4.5. The preceding preorder on triangles is derived from a partial order $>$, on areas (or classes of triangles with the same area).

Partial orders on a finite set must have maximal elements, that is elements such that no other is greater. However, these elements may not be greater than every other.

DEFINITION 1.4.6. Let P be a partial order on a set S. That an element $x \in S$ is *maximal* means if $(x, y) \in P$ then $x = y$. That it is *minimal* means if $(y, x) \in P$ then $x = y$.

EXAMPLE 1.4.6. In the set $\{1, 2, 3\}$ order $\leqslant$, 3 is maximal since $3 \nleqslant x$ for $x \neq 3$, and 1 is minimal.

The names maximal and minimal may be reversed according to the situation. Maximal means $x < y$ is false.

DEFINITION 1.4.7. In a partial order P two elements x, y are *comparable* (*incomparable*) if $(x, y) \in P$ or $(y, x) \in P$ (neither $(x, y) \in P$ nor $(y, x) \in P$).

EXAMPLE 1.4.7. Among subsets of $\{1, 2\}$ the sets $\{1\}$ and $\{2\}$ are not comparable. They form the only incomparable pair.

THEOREM 1.4.3. *Let P be a partial order on a finite set S. Then P has at least one maximal element and at least one minimal element. In fact for $u \in P$ there is at least one maximal element x such that* $(u, x) \in P$.

Proof. For $|S| = 1$, the element is both maximal and minimal. Suppose this theorem holds for $|S| = k$. Let $|S| = k + 1$. Let $u, y \in S$, $u \neq y$. Let Q be P with y removed. By induction assumption Q has a maximal element x, $(u, x) \in P$. If $(x, y) \notin P$ then x is maximal in P also. If $(x, y) \in P$ then y is maximal in P, else if $(y, z) \in P$, $z \neq y$, then $z \in S \backslash \{y\}$, $(x, z) \in Q$ contradicting maximality of x in Q. Moreover by transitivity $(u, y) \in P$. The proof for a minimal element is similar. □

This shows that for a finite partially ordered set (***poset***) any element is $\leqslant$ some maximal element.

EXAMPLE 1.4.8. $\{(1,1), (2, 2), (3, 3), (1, 2), (1, 3)\}$. The set has two maximal elements 2, 3. Neither is greater than the other: they are incomparable.

EXAMPLE 1.4.9. The set of positive integers is not finite, and has no maximal element.

To obtain a result analogous to this theorem for infinite sets, and to further obtain results on partial order, the idea of a ***linear order*** is needed. A *linear order* is a partial order satisfying trichotomy: $x > y$, $y < x$ or $y = x$.

DEFINITION 1.4.8. A binary relation R on a set S is *complete* if for all $x, y \in S$ either $(x, y) \in R$ or $(y, x) \in R$.

EXAMPLE 1.4.10. A partial order which has incomparable elements is not complete. The relation $\leqslant$ on real numbers is complete.

DEFINITION 1.4.9. A *linear* (*total*) *order* is a complete partial order.

EXAMPLE 1.4.11. The relation $\leqslant$ on subsets of the real numbers is a linear order. The relation $\subset$ is not in general.

Linear orders on finite sets S correspond to arrangements of S as $x_1, x_2, \ldots, x_n$, Where $x_i < x_j$ for $i < j$. That is, they have the same structure as $\leqslant$ on the set $\{1, 2. \ldots, n\}$. To state this precisely we need the idea of isomorphism stated earlier.

Two binary operations R_1 on a set S and R_2 on a set T are ***isomorphic*** if and only if there exists a function $h: S \to T$ which is 1-1 and onto, such that $(x_1, x_2) \in R_1$ if and only if $(h(x_1), h(x_2)) \in R_2$.

THEOREM 1.4.4. *Any linear order on a set S of n elements is isomorphic to the linear order* $<$ *on* $\{1, 2, \ldots, n\}$.

Proof. For $n = 1$, this is immediate. Suppose it holds for $n = k$. Let y be a maximal element of S. Let h be an isomorphism of $S\backslash\{y\}$ to $\{1, 2, \ldots, n-1\}$. Set $h(y) = n$. Then h is 1-1 and onto.

Let $u < v$ in S. Then $u \neq y$. If $v \neq y$ then $h(u) < h(v)$ by construction. Suppose $v = y$. Then $h(u) < n$ since n is maximal.

Conversely let $h(u) < h(v)$. If $h(u), h(v) \neq n$ then $u < v$ by construction. Since n is maximal, $u \neq y$. Let $h(v) = n$, i.e. $v = y$, $u \neq v$. Then either $u < y$ or $y < u$. Since y is maximal the latter is false. So $u < v$. □

EXAMPLE 1.4.12. This is not true for infinite linearly ordered sets. The set of negative integers under $<$ is not isomorphic to the set of positive integers under $<$.

DEFINITION 1.4.10. A *chain* in a poset S is a subset which is linearly ordered by the partial order.

EXAMPLE 1.4.13. Among subsets of a set U, a chain is a family of subsets $S_1, S_2, \ldots, S_k$ with $S_1 \subset S_2 \subset \ldots \subset S_k$.

Finally the generalization of the result about maximal elements can be stated (but not proved).

Hausdorff's Maximal Principle. *Every poset has a maximal chain.*

This cannot be proved without the use of a special axiom of set theory, the axiom of choice.

AXIOM OF CHOICE. For any indexed family F_α of nonempty sets there exists an indexed family x_α such that $x_\alpha \in F_\alpha$ for each α.

The Hausdorff maximal principle and its equivalents are quite important in abstract algebra, for infinite sets. An example is as follows.

THEOREM 1.4.5. *Every partial order on a set S is contained in a linear order on S.*

Proof. Let P be a partial order on S. Consider the family of all partial orders on S containing P. It is itself a poset under inclusion. Take a maximal chain C in that set. The union U of its members will also be a partial order on S. For example take $x, y, z \in S$. Let $(x, y) \in U$, $(y, z) \in U$. Then $(x, y) \in P_1$, $(y, z) \in P_2$ for some P_1, P_2 in C. By definition of chain, $P_1 \subset P_2$ or $P_2 \subset P_1$. Assume the latter. Then (x, y), $(y, z) \in P_1$ so $(x, z) \in P_1$. So $(x, z) \in U$. This proves U is transitive. Proofs of reflexivity and antisymmetry are similar. Since C is maximal, U is not contained in any other partial order on S else that would give a larger chain. This will be used to get a contradiction if U is not a linear order. Suppose not. Then for some $x \neq y$, $(x, y) \notin U$, $(y, x) \notin U$. Then let $T = U$ together with $\{(w, z)\colon (w, x) \in U \text{ and } (y, z) \in U\}$. Then T properly contains U since $(x, y) \in T$. It can be shown that T is a partial order, by checking various cases. For example let $(w, z) \in T\backslash U$, $(z, s) \in U$. Then $(w, x) \in U$, $(y, z) \in U$. Then $(w, x) \in U$, $(y, s) \in U$. So $(w, s) \in T$. Let $(w_1, z_1) \in T\backslash U$, $(z_1, z_2) \in T\backslash U$. Then $(y, z_1) \in U$, $(z_1, x) \in U$ so $(y, x) \in U$. This is false. The other cases of transitivity and antisymmetry are proved in a similar way. This gives a contradiction. □

The following consequence of the Axiom of Choice is called ***Zorn's Lemma.*** *Suppose P is a poset and every chain has an upper bound. Then P has a maximal element.*

EXERCISES

Level 1

1. Prove that the identity relation on any set is a partial order. Is any equivalence relation a preorder (quasiorder)?
2. What are the maximal and minimal elements of this strict partial order: $\{(1, 3), (1, 4), (2, 3), (2, 4)\}$? There are two of each.
3. What are the maximal and minimal elements of this strict partial order: $\{(1, 3), (2, 3)\}$?
4. Finite posets are represented by diagrams called *Hasse diagrams*. To draw one, first find the minimal elements and label them as separate points at the bottom level. Then for all points z such that $(m, z) \in P$ for minimal m but

there is no intermediate y, with $(m, y) \in P$ and $(y, z) \in P$ draw z on a level above m and a line from z to m. Then draw the elements w such that $(z, w) \in P$ but there is no intermediate y. For example: in $\{(1, 2), (1, 3)\}$, 1 is minimal. Both 2 and 3 are above it. So we have

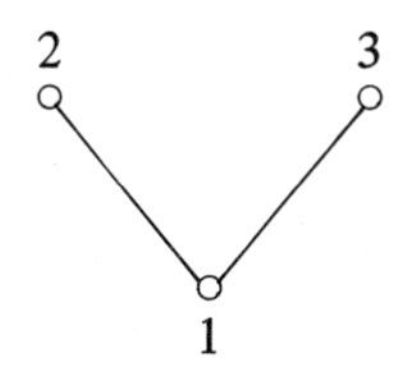

Do this for Exercise 2.

5. Draw the Hasse diagram for Exercise 3.
6. Find 3 partial orders on $\{1, 2\}$ (include the identity).
7. If P_1 is a partial order on S_1, P_2 is a partial order on S_2 and $S_1 \cap S_2 = \emptyset$, show $P_1 \cup P_2$ is a partial order on $S_1 \cup S_2$.
8. An example of exercise 7 is $P_1 = \{(1, 2)\}$, $P_2 = \{(3, 4)\}$. Draw the Hasse diagram of $P_1 \cup P_2$.

Level 2

1. Draw the Hasse diagram of $\{(1, 2), (2, 3), (1, 3)\}$.
2. Draw all possible Hasse diagrams for 3 elements (there are 5). Two are as follows:

3. What is the diagram of any identity relation? any linear order?
4. Label the subsets of $\{1, 2, 3\}$ as $\{a, b, \ldots, h\}$. Write out the strict partial order for inclusion of proper subsets. Draw the Hasse diagram. (It should look like a cube drawn on the plane. There is one minimal element $\emptyset$ and one maximal element $\{1, 2, 3\}$.)
5. Prove any partial order P on a set S is isomorphic to a subset of the poset of subsets of S. Send $x \in S$ to $h(x) = \{y\colon (y, x) \in P\}$. Show that $(x, y) \in P$ if and only if $h(x) \subset h(y)$. This implies h is 1–1.
6. Write out all preorders on $\{1, 2, 3\}$ such that $1\ P\ 2$.
7. If P_1 is a partial order on S_1 and P_2 is a partial order on S_2 show that there is a corresponding partial order $P_1 \times P_2$ on $S_1 \times S_2$.

Level 3

1. Draw all possible Hasse diagrams (19) for 4 elements.
2. Prove that every partial order P on a set S is an intersection of linear orders on S. It suffices to show that if $(y, x) \notin P$ then there exists a linear

order L such that $(y, x) \notin L$ but $P \subset L$. (P will then be the intersection of all these L.) The proof is similar to the proof of Theorem 1.4.5, with C being the family of all partial orders containing P but not (y, x). The same T will do.

3. Represent the strict partial order $\{(1, 2), (1, 3)\}$ as an intersection of linear orders.
4. A ***semiorder*** is a binary relation R such that (1) R is irreflexive, (2) if $(x, y) \in R$ and $(z, w) \in R$ then $(x, w) \in R$ or $(z, y) \in R$, (3) if $(x, y) \in R$ and $(y, z) \in R$ then for all w either $(x, w) \in R$ or $(w, z) \in R$. Prove a semiorder is transitive, and is therefore a strict partial order.
5. Every 3-element partial order is a semiorder. Find two 4-element partial orders which respectively violate (2), (3) of the above definition and are not semiorders.
6. Suppose S is a finite set and f a function $f\colon S \to T$ and $\epsilon > 0$. Then show $\{(x, y)\colon f(y) > f(x) + \epsilon\}$ is a semiorder. This leads to the interpretation that 'x is distinguishably greater than y' is a semiorder. The converse of this result is also true. See Scott and Suppes (1958), Luce (1956), Rabinovitch (1977), Scott (1964), Suppes and Zinnes (1963).
7. Prove for a semiorder R that the sets $L_x^1 = \{y\colon (x, y) \in P\}$ and the sets $L_x^2 = \{y\colon (y, x) \in P\}$ are each linearly ordered under inclusion and that $L_x^1 \supsetneqq L_y^1$, $L_x^2 \supsetneqq L_y^2$ cannot happen.

1.5 BOOLEAN MATRICES AND GRAPHS

The simplest ***Boolean algebra*** is $\mathcal{B} = \{0, 1\}$. Its operations are given by

$$0 + 0 = 0 \cdot 1 = 1 \cdot 0 = 0 \cdot 0 = 0$$

$$1 \cdot 1 = 1 + 0 = 0 + 1 = 1 + 1 = 1$$

(except for $1 + 1$, the same as for ordinary addition and multiplication.) However, it has this difference, $\mathcal{B}$ obeys the rules of set intersection and union, rather than those of arithmetic. For example,

$$x + y = y + x,\ xy = yx,\ x + (y + z) = (x + y) + z,\ x(yz) = x(yz)$$

$$x(y + z) = xy + xz,\ x + yz = (x + y)(x + z) \qquad \text{(Dual Distributive)}$$

$$x + 0 = x + x = x,\ x \cdot 1 = x \cdot x = x,\ x \cdot 0 = 0,\ x \cdot 1 = x$$

There is also the operation x^{C}, ***complement*** such that $1^{\mathrm{C}} = 0$, $0^{\mathrm{C}} = 1$.

$$(x^{\mathrm{C}})^{\mathrm{C}} = x,\ \ (x + y)^{\mathrm{C}} = x^{\mathrm{C}}y^{\mathrm{C}},\ \ (xy)^{\mathrm{C}} = x^{\mathrm{C}} + y^{\mathrm{C}}$$

A partial order is defined by $x \leqslant x$, $0 \leqslant 1$. If $x \leqslant y$ then $x + z \leqslant y + z$, $xz \leqslant yz$, and $x^{\mathrm{C}} \leqslant y^{\mathrm{C}}$.

Any system with operations denoted by addition, multiplication, and complementation, satisfying the laws just given, is called a *Boolean algebra.* Boolean algebras have a number of applications.

EXAMPLE 1.5.1. Logic can be studied using $\mathcal{B}$. In this case, '1' is 'true', '0' is 'false', '+' is 'or', '×' is 'and', 'C' is 'not'. So (p or q) and not r would be translated

$$(p + q)r^C$$

Statement x implies y if and only if $x \leqslant y$.

EXAMPLE 1.5.2. In mathematics, in dealing with nonnegative numbers, 0 can represent 0 and 1 can represent any positive number. Then $1 + 1 = 1$ refers to the fact that the sum of any two positive numbers is positive.

EXAMPLE 1.5.3. Sets can be represented by n-tuples $(0, 1, 1, \ldots, 0)$ of Boolean numbers. A one in place i means that $x_i \in S$ and a zero in place i means $x_i \notin S$. Then sum is union, product is intersection, complement is complement.

Boolean algebras are also used in the study of computer and other switching circuits.

DEFINITION 1.5.1. An $n \times m$ *Boolean matrix* M is an $n \times m$ array of numbers m_{ij} from $\mathcal{B} = \{0, 1\}$. The number m_{ij} lies in the ith row (horizontal line) and jth column (vertical line) of M. It is called the *(i, j)-entry.*

Matrices over a field will be treated in section 3.3.

EXAMPLE 1.5.4. This is a Boolean matrix M, where $m_{12} = 1$, $m_{21} = 0$.

$$\begin{bmatrix} 0 & 1 & 0 \\ 0 & 1 & 0 \end{bmatrix}$$

The following operations are simply taken entry by entry for Boolean matrices $A = (a_{ij})$, $B = (b_{ij})$.

Sum: $A + B = (a_{ij} + b_{ij})$

Logical Product: $A \odot B = (a_{ij} b_{ij})$

Complement: $A^C = (a_{ij}{}^C)$

Inequality is also defined entrywise: $A \leqslant B$ if and only if for all i, j, $a_{ij} \leqslant b_{ij}$. This is a partial order. The strict partial order $A < B$ means $a_{ij} < b_{ij}$ for at least one i, j and $a_{ij} \leqslant b_{ij}$ for all i, j.

EXAMPLE 1.5.5.

(a) $$\begin{bmatrix} 1 & 0 \\ 1 & 0 \end{bmatrix} + \begin{bmatrix} 1 & 1 \\ 0 & 0 \end{bmatrix} = \begin{bmatrix} 1 & 1 \\ 1 & 0 \end{bmatrix}$$

(b) $$\begin{bmatrix} 1 & 0 \\ 1 & 0 \end{bmatrix} \odot \begin{bmatrix} 1 & 1 \\ 0 & 0 \end{bmatrix} = \begin{bmatrix} 1 & 0 \\ 0 & 0 \end{bmatrix}$$

(c) $$\begin{bmatrix} 1 & 0 \\ 0 & 0 \end{bmatrix}^{C} = \begin{bmatrix} 0 & 1 \\ 1 & 1 \end{bmatrix}$$

(d) $$\begin{bmatrix} 1 & 0 \\ 0 & 1 \end{bmatrix} < \begin{bmatrix} 1 & 1 \\ 0 & 1 \end{bmatrix}$$

Inequality means the places where 1 occurs in one matrix is a subset of the places where 0 occurs. The operation $+, \odot$, C, and the relation $\leqslant$ obey all the laws stated earlier for $\mathcal{B}$.

There are two other operations on Boolean matrices: ***transposes*** and ***matrix multiplication***.

$$\text{Transpose:} \qquad A^{\mathrm{T}} = (a_{ji})$$

$$\text{Matrix Product:} \; AB = (\Sigma\, a_{ij} b_{jk})$$

EXAMPLE 1.5.6.

(a) $$\begin{bmatrix} 0 & 1 \\ 0 & 0 \end{bmatrix}^{\mathrm{T}} = \begin{bmatrix} 0 & 0 \\ 1 & 0 \end{bmatrix}$$

(b) $$\begin{bmatrix} 1 & 0 \\ 1 & 1 \end{bmatrix} \begin{bmatrix} 1 & 1 \\ 0 & 1 \end{bmatrix} = \begin{bmatrix} 1 \cdot 1 + 0 \cdot 0 & 1 \cdot 1 + 0 \cdot 1 \\ 1 \cdot 1 + 1 \cdot 0 & 1 \cdot 1 + 1 \cdot 1 \end{bmatrix} = \begin{bmatrix} 1 & 1 \\ 1 & 1 \end{bmatrix}$$

(c) $$\begin{bmatrix} 0 & 0 & 1 \\ 1 & 0 & 0 \\ 0 & 1 & 1 \end{bmatrix} \begin{bmatrix} 0 & 1 & 0 \\ 0 & 0 & 1 \\ 1 & 1 & 1 \end{bmatrix} = \begin{bmatrix} 1 & 1 & 1 \\ 0 & 1 & 0 \\ 1 & 1 & 1 \end{bmatrix}$$

The transpose changes the (i, j)-entry to location j, i. Rows of A become columns of A^{T}, and vice versa.

The matrix product is the same as the product of ordinary matrices except that operations are Boolean, so that $1 + 1 = 1$. The (i, j)-entry may be calculated by multiplying the entries of row i of A by those of column j of B, and adding the sums. Products are defined only if the number of columns of A equals the number of rows of B. An $n \times m$ matrix times an $m \times s$ matrix gives an $n \times s$ matrix.

There is also a simpler way to multiply Boolean matrices. To form row i of the product, consider the ones in row i of A. If these are in locations $i_1, i_2, \ldots, i_k$, add rows $i_1, i_2, \ldots, i_k$ of B. This gives row i of the product.

The *ith row* (*column*) of A is denoted A_{i*} (A_{*i}).

EXAMPLE 1.5.7. To form the product of Example 1.5.6. (c), A_{1*}: $a_{13} = 1$ take $B_{3*} = (1\ 1\ 1)$. A_{2*}: $a_{21} = 1$ take $B_{1*} = (0\ 1\ 0)$. A_{3*}: $a_{32} = a_{33} = 1$ take $B_{2*} + B_{3*} = (0\ 0\ 1) + (1\ 1\ 1) = (1\ 1\ 1)$.

Transpose obeys these laws:

$$(A+B)^{\mathrm{T}} = A^{\mathrm{T}} + B^{\mathrm{T}}$$

$$(A \odot B)^{\mathrm{T}} = A^{\mathrm{T}} \odot B^{\mathrm{T}}$$

$$(A^{\mathrm{C}})^{\mathrm{T}} = (A^{\mathrm{T}})^{\mathrm{C}}$$

$$(AB)^{\mathrm{T}} = B^{\mathrm{T}}A^{\mathrm{T}}$$

$$A \leqslant B \Rightarrow A^{\mathrm{T}} \leqslant B^{\mathrm{T}}$$

Boolean matrix products obey these laws (as ordinary nonnegative matrices do):

$$(AB)C = (AB)C$$

$$(A+B)C = AC + BC, \quad C(A+B) = CA + CB$$

$$A \leqslant B \Rightarrow AC \leqslant BC, \quad CA \leqslant CB$$

In addition to its relation to multiplication of nonnegative matrices, Boolean matrix multiplication is used because it corresponds to composition of binary relations.

THEOREM 1.5.1. *Let S, T be a sets of n, m elements. Label their elements $x_1, x_2, \ldots, x_n$, $y_1, y_2, \ldots, y_m$. Associate a Boolean matrix A to each binary relation R by the rule $a_{ij} = 1$ if and only if $(x_i, y_j) \in R$. This gives a 1-1 correspondence between relations on S, T and $n \times m$ Boolean matrices. Under this correspondence unions become sums, intersections logical products, complements complements, compositions matrix products, inclusions become $\leqslant$, and 'converse' sending (x, y) to (y, x) becomes transpose.*

Proof. The proof is straightforward but lengthy. The matrix entries determine precisely which pairs are in R. And for any matrix A we can define a relation by

$\{(x_i, x_j): a_{ij} = 1\}$ which goes to it. This proves the mapping of relations to matrices is 1-1 and onto.

We verify the statement about matrix products. Take a third set $Z = \{z_1, z_2, \dots, z_q\}$ and a second relation P with matrix B. Let $C = AB$. Then $c_{ij} = 1$ if and only if $\Sigma a_{ik} b_{kj} = 1$, i.e., for some k, $a_{ik} b_{kj} = 1$ if and only if for some k, $a_{ik} = 1$ and $b_{kj} = 1$ if and only if for some k, $(x_i, y_k) \in R$ and $(y_k, z_j) \in P$ if and only if $(x_i, z_j) \in R \circ P$.

Proofs for other operations are similar. □

For a comprehensive treatment of Boolean matrix theory, see Kim (1982).

Here we will view graphs as representations of binary relations. A ***directed graph*** (not like the graph of a function) consists of points called *vertices*, and line segments with arrows from certain vertices to certain others. The line segments are called ***directed edges***. The location of points or the shape of edges do not affect the structure of a directed graph.

A directed graph represents a binary relation R on a set S if its vertices are labelled to correspond to the elements of S and a directed edge is drawn from x to y if and only if $(x, y) \in R$.

EXAMPLE 1.5.8. The relation $\{(1, 1), (1, 2), (2, 2)\}$ has this graph.

EXAMPLE 1.5.9. The relation $\{(1, 5), (2, 3), (3, 4), (2, 4)\}$ has this graph.

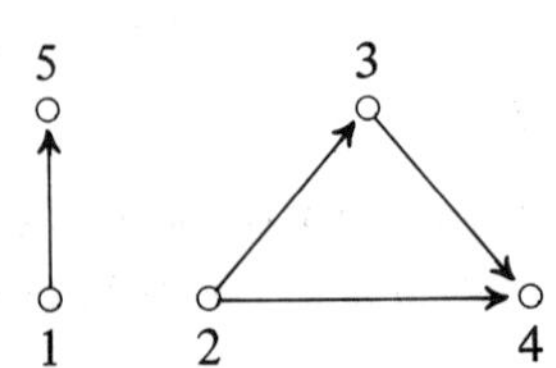

First the correct number of points should be drawn. Then they can be labelled. Then for each pair in the binary relation, a directed segment is drawn between the vertices having the same labels. Two-way arrows (or edges without arrows) may be used if (x, y) and (y, x) are in R.

Graphs are useful in studying questions related to powers of a binary relation under composition $R \circ R \circ \dots \circ R$. In graph theory questions of connectedness, existence of various types of paths, and families of subsets such that no point joins another of the same subset are studied.

The graph of an $n \times n$ Boolean matrix can be drawn. Vertices are drawn and labelled $1, 2, \dots, n$. An arrow is drawn from i to j for each entry a_{ij} which is 1.

EXERCISES

Level 1

1. Compute $(1 + 1^{\mathrm{C}}) \cdot (0 + 0^{\mathrm{C}})$ in Boolean arithmetic.
2. Prove the dual distributive law $x + yz = (x + y)(x + z)$ in Boolean algebra by considering three cases (1) $x = 1$, (2) $y = z = 1$, (3) $x = 0$ and y or z is 0.
3. When is xy equal to 1? How is this like a statement 'x and y'?
4. Add

$$\begin{bmatrix} 1 & 1 & 0 \\ 0 & 1 & 1 \\ 1 & 1 & 0 \end{bmatrix} + \begin{bmatrix} 0 & 0 & 1 \\ 0 & 1 & 0 \\ 1 & 0 & 0 \end{bmatrix}$$

5. Find the logical product

$$\begin{bmatrix} 1 & 1 & 0 \\ 0 & 1 & 1 \\ 1 & 1 & 0 \end{bmatrix} \odot \begin{bmatrix} 0 & 0 & 1 \\ 0 & 1 & 0 \\ 1 & 0 & 0 \end{bmatrix}$$

6. Draw the graph of the Boolean matrix

$$\begin{bmatrix} 0 & 1 & 0 \\ 1 & 0 & 1 \\ 0 & 1 & 0 \end{bmatrix}$$

Level 2

1. Prove $x + y = x$ if and only if $x \geqslant y$ in a Boolean algebra.
2. Prove addition of Boolean matrices is associative (assuming addition is associative in $\mathcal{B}$).
3. Prove the distributive law for Boolean matrix products (assuming the laws for $\mathcal{B}$).
4. Multiply

$$\begin{bmatrix} 1 & 0 & 1 & 0 \\ 1 & 0 & 1 & 0 \\ 0 & 1 & 0 & 1 \\ 0 & 1 & 0 & 1 \end{bmatrix}$$

times itself.

5. Find all distinct powers of

$$\begin{bmatrix} 0 & 1 & 0 & 0 \\ 0 & 0 & 1 & 0 \\ 0 & 0 & 0 & 1 \\ 1 & 1 & 0 & 0 \end{bmatrix}$$

6. Let J be the Boolean matrix all of whose entries are 1. When is $MJ = J$?
7. Let I be the $n \times n$ ***identity matrix*** (δ_{ij}) where $\delta_{ij} = 0$ if $i \neq j$ and $\delta_{ii} = 1$ for all i. Prove $AI = IA = A$ for every $n \times n$ Boolean matrix A.
8. Prove a relation is transitive if and only if its Boolean matrix satisfies $A^2 \leqslant A$. Show that $A^2 = A$ if it is reflexive also.
9. For a preorder Q, with Boolean matrix A, show $A \odot A^{\mathrm{T}}$ is the Boolean matrix of the indifference relation Q_{D}.

Level 3

1. Suppose $\mathcal{B}$ is replaced by any system in which $+, \times$ are defined. Can Boolean matrix multiplication be defined?
2. A ***Boolean row vector*** v is a $1 \times n$ matrix. Prove that $A = B$ if and only if $vA = vB$ for all row vectors v. (Choose v with 1 entry.)
3. Prove $(A_{i*})B$ is the ith row of AB.
4. Prove that if $I \leqslant A$, $v \leqslant vA$ and that if $v = vA$ then $v = vA^i$ for all $i > 0$. Here I is an identity matrix.
5. Use (2), (3), (4) to prove that for $A \geqslant I$, A^{n-1} equals all powers $A^k, k > n - 1$.
6. A ***row basis*** is the set of vectors formed by rows of A which are not sums of other, lesser rows of A. The **row rank** ρ_{r} of a Boolean matrix is the number of vectors in a row basis. Prove that $\rho_{\mathrm{r}}(AB) \leqslant \rho_{\mathrm{r}}(A)$. Must $\rho_{\mathrm{r}}(AB) \leqslant \rho_{\mathrm{r}}(B)$? Try 4×4 Boolean matrices with 3 nonzero rows.
7. Prove that in a graph every point is accessible from every other by a directed sequence of edges if and only if $(I + A)^n$ has all ones in it.
8. Prove $(I + A)^n$ is the Boolean matrix of a preorder. This relation is called *reachability*. What is its graph-theoretic interpretation?

1.6 ARROW'S IMPOSSIBILITY THEOREM

Majority rule is a traditional method of group decision-making in the West. In fact it can be proved to be the only rule which satisfies certain conditions like giving every person and every alternative equal weight, and favouring an alternative more as more people prefer it over another choice.

For three or more alternatives, a problem known as the ***Condorcet paradox*** can occur. Suppose there are three voters 1, 2, 3 and three alternatives *a, b, c*. Let the voter preference orderings be *abc* (meaning *a* is preferred to *b* and *b* to *c*), *bca, cab* for voters 1, 2, 3. Then two voters prefer *a* to *b* (voters 1 and 3).

Also two voters prefer b to c and two voters prefer c to a. Thus for any chosen alternative, a majority will prefer some other alternative.

Can a more complicated voting procedure avoid this paradox? Kenneth Arrow in 1950 proved there is no voting procedure satisfying five basic conditions. (However, voting procedures can exist satisfying all but one, called *independence of irrelevant alternatives*.)

Let X be a set of alternatives which the group N of voters, represented as $\{1, 2 \ldots, n\}$ will choose. The preferences of individual i will be expressed as a binary relation R_i where $x R_i y$ is interpreted as the voter liking x at least as well as y, for $x, y \in X$. If the voter is rational, the relation should be transitive: if he likes y at least as well as x, and z at least as well as y, he will like z at least as well as y. Moreover it should be complete: either he likes x at least as well as y or he likes y better than y.

DEFINITION 1.6.1. A *weak order* is complete, transitive binary relation. Thus it is a ***total preorder***.

EXAMPLE 1.6.1. A linear order is a weak order.

Here we will assume that no voter is exactly indifferent between any two alternatives, so that R_i is a linear order.

For the linear order R_i of preferences of voter i, let P_i denote the corresponding strict partial order.

The group preference is expressed as a social welfare function $\mathbb{F}(R_1, R_2, \ldots, R_n)$. Here $\mathbb{F}$ is the function which assigns to the individual preference relations $R_1, R_2, \ldots, R_n$ the group preference relation in this case. The group preference relation gives the set of pairs (x, y) of alternatives such that group either chooses x over y or is indifferent about them. The binary relations $R_1, R_2, \ldots, R_n$ lie in a set of binary relations called the ***domain***.

DEFINITION 1.6.2. A *social welfare function* is a function $\mathbb{F}$ from a set $\mathbb{D} \subset W^n = W \times W \times \ldots \times W$ to $\mathcal{B}_X$. Here $\mathbb{D}$ is the domain, W is the set of weak orders on X, W^n is its n-fold Cartesian product with itself, and $\mathcal{B}_X$ is the set of binary relations on X.

Frequently we write $\mathbb{F}$ for $\mathbb{F}(R_1, R_2, \ldots, R_n)$.

EXAMPLE 1.6.2. ***Majority rule*** is a social welfare function, where $x \mathbb{F} y$ if and only if $|\{i : x R_i y\}| > n/2$. This is a rule by which the individual preferences R_i determine a group preference $\mathbb{F}$. Any such definite rule can be expressed as a social welfare function.

Both the value of the function $\mathbb{F}$ and the individual preference relation R_i will be represented by Boolean matrices, as was done in the last section. So a fixed labelling of the elements of X as $x_1, x_2, \ldots, x_m$ will be used. Then the Boolean matrix $R\langle i\rangle$ of R_i will be 1 in (j, k)-entry if $x_j R\langle i\rangle x_k$, else it will be zero.

The first two assumptions on the social welfare function are universal domain and weak ordering.

ASSUMPTION 1. UNIVERSAL DOMAIN. The domain $\mathbb{D}$ *is the set of all linear orders on* X.

Thus every individual preference which is a linear order can occur (weak orders are frequently used instead).

ASSUMPTION 2. WEAK ORDERING. The relation $\mathbb{F}(R_1, R_2, \ldots, R_n)$ *expressing group preferences is a weak order.*

PROPOSITION 1.6.1. *A Boolean matrix A is the matrix of a weak order if and only if for all i, j, k, (1)* $a_{ij} + a_{ji} = 1$, *(2)* $a_{ij}\, a_{jk} \leqslant a_{ik}$.

Proof. It can be checked that the first condition is equivalent to completeness ($x_i \mathrm{R}\, x_j$ or $x_j \mathrm{R}\, x_i$) and the second to transitivity (if $x_i \mathrm{R}\, x_j$ and $x_j \mathrm{R}\, x_k$ then $x_i \mathrm{R}\, x_k$). □

The third condition is ***Pareto optimality***. This means if every individual strictly prefers x to y, the group strictly prefers x to y.

DEFINITION 1.6.3. That a social welfare function $\mathbb{F}$ is *Pareto optimal* means if $x\, P_i\, y$ for all i then $x\, \mathbb{F}(R_1, R_2, \ldots, R_n)\, y$ but not $y\, \mathbb{F}(R_1, R_2, \ldots, R_n)\, x$.

EXAMPLE 1.6.3. Majority rule satisfies this condition: if every one prefers x to y then a majority does.

PROPOSITION 1.6.2. *A social welfare function* $\mathbb{F}$ *on linear orders is Pareto optimal if and only if* $R\langle 1\rangle \odot R\langle 2\rangle \odot \ldots \odot R\langle n\rangle \leqslant \mathbb{F}(R_1, R_2, \ldots, R_n) \leqslant R\langle 1\rangle + R\langle 2\rangle + \ldots + R\langle n\rangle$ *in Boolean matrix terms.*

Proof. Pareto optimality means that if the (i, j)-entries of $R\langle i\rangle$ are 1 and the (j, i)-entries are zero (each of which implies the other for linear orders), then $f_{ij} = 1$, $f_{ji} = 0$. So if all R_i have a 1 entry $\mathbb{F}$ is 1. This is $\mathbb{F} \geqslant R\langle 1\rangle \odot R\langle 2\rangle \odot \ldots \odot R\langle n\rangle$ in the (i, j)-entry. And if all $R\langle i\rangle$ have a 0 entry, $\mathbb{F}$ is 0. This is $\mathbb{F} \leqslant R\langle 1\rangle + R\langle 2\rangle + \ldots + R\langle n\rangle$. □

The fourth condition is ***nondictatoriality***. This means there is no fixed individual k such that the group preferences are always identical with his.

DEFINITION 1.6.4. A social welfare function $\mathbb{F}$ is *dictatorial* if for some k, $\mathbb{F}(R_1, R_2, \ldots, R_n) = R_k$.

EXAMPLE 1.6.4. $\mathbb{F} = R_1$ is a dictatorial social welfare function.

The fifth condition is ***independence of irrelevant alternatives***. This means the group preference between alternatives x, y should not be affected by how the group feels about some third alternative z. While this might ideally be true there is a question as to whether it is necessary for practical voting methods, for which preferences about z may reveal something about intensities of preferences for x, y. If the alternative set is very large, or not precisely defined, though, it may also be a practical necessity.

DEFINITION 1.6.5. That a social welfare function $\mathbb{F}$ is *independent of irrelevant alternatives* means that if for some x, y and preferences R_i, Q_i, if $x R_i y$ if and only if $x Q_i y$ and $y R_i x$ if and only if $y Q_i x$ then $x \mathbb{F}(R_1, R_2, \ldots, R_n) y$ if and only if $x \mathbb{F}(Q_1, Q_2, \ldots, Q_n) y$.

EXAMPLE 1.6.5. Majority voting and any dictatorial social welfare function are independent of irrelevant alternatives.

This condition means that if preferences of each individual i over x, y are the same in profiles R, Q (a ***profile*** is an n-tuple of preferences) then the group preference is the same in R, Q regarding x, y. The remaining assumptions are these.

ASSUMPTION 3. PARETO OPTIMALITY. $\mathbb{F}$ is Pareto optimal.

ASSUMPTION 4. NONDICTATORIALITY. $\mathbb{F}$ is nondictatorial.

ASSUMPTION 5. INDEPENDENCE OF IRRELEVANT ALTERNATIVES. $\mathbb{F}$ is independent of irrelevant alternatives.

PROPOSITION 1.6.3. *For a social welfare function $\mathbb{F}$ which is independent of irrelevant alternatives, f_{jk} is a function of $r_{jk}\langle i \rangle$.*

Proof. Being a function means, that if for all i, $r_{jk}\langle i \rangle = q_{jk}\langle i \rangle$ then $f_{jk}(R_1, R_2, \ldots, R_n) = f_{jk}(Q_1, Q_2, \ldots, Q_n)$. But for linear orders, $r_{jk} = r_{kj}{}^{\mathrm{C}}$ so $r_{kj}\langle i \rangle = q_{kj}\langle i \rangle$ also. Now the assertion follows directly from the definition. □

THEOREM 1.6.4. (***Arrow's Impossibility Theorem***). *If X contains at least three alternatives, there exists no social welfare function satisfying assumptions (1) to (5).*

Proof. Write $f_{ij}(a_1, a_2, \ldots, a_n)$ to express f_{ij} as a function of the (i, j)-entries of $R_1, R_2, \ldots, R_n$ by Proposition 1.6.3. By completeness for $i \neq j$,

$$f_{ij}(a_1, a_2, \ldots, a_n) + f_{ji}(a_1^{\mathrm{C}}, a_2^{\mathrm{C}}, \ldots, a_n^{\mathrm{C}}) = 1$$

By the Pareto condition

$$f_{ij}(1, 1, \ldots, 1) = 1$$
$$f_{ij}(0, 0, \ldots, 0) = 0$$

Let $a_1, a_2, \ldots, a_n, b_1, b_2, \ldots, b_n, c_1, c_2, \ldots, c_n$ be such that $a_i b_i \leqslant c_i \leqslant a_i + b_i$ for all i. We will construct a profile R_i such that for any three fixed alternatives x, y, z, $x\,R_i\,y$ if and only if $a_i = 1$, $y\,R_i\,z$ if and only if $b_i = 1$, $x\,R_i\,z$ if and only if $c_i = 1$. To do this use this table:

a_i	b_i	c_i	Preference
0	0	0	zyx
1	0	0	zxy
0	1	0	yzx
1	0	1	xzy
0	1	1	yxz
1	1	1	xyz

This proves the claim. Thus by transitivity and universal domain, if $a_i b_i \leqslant c_i \leqslant a_i + b_i$ then $f_{ij}(a) f_{jk}(b) \leqslant f_{ik}(c)$ where $a = (a_1, a_2, \ldots, a_n)$, and so on. Take $a = c$, $b = (1, 1, \ldots, 1)$. Then $f_{ij}(a) \leqslant f_{ik}(a)$. By symmetry $f_{ij}(a) = f_{ik}(a)$. By a dual argument $f_{ji}(a) = f_{ki}(a)$. Since this is true for all distinct i, j, k we have that all f_{ij} for $i \neq j$, are equal (note $f_{ii}(a) = 1$ identically by completeness). Let f denote this common function.

For $a_i b_i \leqslant c_i \leqslant a_i + b_i$, $f(a) f(b) \leqslant f(c)$. Suppose $f(a) = 1$ and $c_i \geqslant a_i$. Let $b_i = 1$. We have $f(c) \geqslant 1$ so $f(c) = 1$. So f is monotone. Choose a vector w with the fewest number of ones such that $f(w) = 1$. Suppose $f(v) = 1$ for any v. Then let $c_i = v_i w_i$. Then $f(c) \geqslant 1 \cdot 1 = 1$. But unless $v > w$, c will have fewer ones than w. This is impossible. So only vectors greater than w can have $f(v) = 1$.

By Pareto optimality, $w \neq c$. Suppose w has two or more ones. Let $0 < v < w$. Then $f(v) = 0$ since $v \not> w$ and $f(v^{\mathrm{C}}) = 0$ since $v^{\mathrm{C}} \not> w$. This contradicts completeness. So w has exactly one 1, say in place t.

Then $f(v) = 1$ if and only if $v \geqslant w$, if and only if $v_t = 1$. So $x\,|\!\mathrm{F}\,y$ if and only if $x\,R_t\,y$. So voter t is a dictator. □

EXERCISES

Level 1

1. For only two alternatives, name a social welfare function satisfying assumptions 1–5.
2. What assumption does majority rule fail to satisfy in general?
3. Which assumptions does a dictatorial social welfare function satisfy?
4. Suppose we count point totals as follows: an alternative receives $m-1$ points for being first choice of anyone, $m-2$ for being second, etc., where

$m = |X|$. Is this transitive? complete? Pareto optimal? nondictatorial? Can it satisfy universal domain? Here $x \, \mathbb{F} \, y$ if and only if x receives a greater point total than y.

5. Show the function in Exercise 4 is not independent of irrelevant alternatives. Take profiles P_1 and P_2 with

P_1	P_2
xyz	xyz
yzx	yxz
zxy	xyz

6. Which assumption is least necessary for real applications of majority rule? How does it really work despite violating the five assumptions?

Level 2

1. List all weak orders on the set $\{1, 2, 3\}$. A weak order is a ranking of certain alternatives as first, others as second, others as third, where ties are allowed.
2. Prove that majority rule is transitive on any domain of the form L^n where L is a set of linear orders such that for no three alternatives x, y, z do we have three orderings xyz, yzx, zxy? (This is called a *cyclic triple*.) Suppose not. Suppose a majority prefers x to y and a majority prefers y to z but a majority does not prefer x to z. Prove at least one voter prefers x to y and y to z, at least one prefers z to x and x to y, at least one prefers y to z and z to x. This is a contradiction.
3. A set of preferences is called *single peaked* if the alternatives can be ordered $x_1, x_2, \ldots, x_n$ so that no one ranks x_j below both x_i, x_k if $i < j < k$. Prove that no cyclic triple can occur for single peaked preferences. (Thus majority rule is transitive.) Suppose xyz, yzx, zxy are single peaked preferences. Let x be the middle numbered one of x, y, z. Can yzx occur?
4. Graphically being single peaked means that as one goes from left to right along $x_1, x_2, \ldots, x_n$ the utility (value) of x_i to a person rises to a peak and then falls. Explain why this may be true in an election where x_1 is the most conservative candidate, x_m the most liberal and the others are arranged in order of liberalness or conservativeness.
5. Give an example of a social welfare function satisfying all the assumptions except Pareto optimality. (It could be a constant function.)

Level 3

1. Prove Arrow's Impossibility Theorem for the domain $\mathbb{D}$ of all weak orders (nondictatorial must be weakened so that just if voter k strictly prefers any x to y so does the group. If he is indifferent, the group choice may differ).
2. Explain why the number of weak orders on a set of m elements is $\Sigma\, S(n, k)k!$. Use the classification of preorders. The partial order involved must be a weak order.

3. Suppose the assumption of completeness is dropped from Arrow's Impossibility Theorem. What can be concluded?
4. A nonempty family F of subsets of a finite set is called a *filter* (1) if $A \in F$ and $B \supset A$ then $B \in F$, (2) if $A, B \in F$ then $A \cap B \in F$, (3) $\emptyset \notin F$. Classify filters on a finite set.
5. Find a different type of filter on an infinite set. A filter is called an *ultra-filter* if for all A either $A \in F$ or $A^C \in F$. Show by Zorn's Lemma that any filter is a subfamily of an ultrafilter.

CHAPTER 2

Semigroups and groups

In this chapter we take up systems having a single binary operation, such as multiplication. Most systems of interest satisfy the associative property $(ab)c = a(bc)$ or a related property. In them the commutative property $ba = ab$ is not always true (for instance matrix multiplication). A system with the associative property is called a ***semigroup***. Semigroups can be studied in several ways. One is to find generators and relations. A set of elements is a set of generators if all elements of the system are products of a sequence from the set. For instance $-1, 1$ generate the integers as a semigroup under addition since every interger is a sum of copies of $-1, 1$. Relations tell when two products of generators are the same, for instance $1 + 1 - 1 = 1$.

Another method is to study subsemigroups, ideals, and quotient semigroups. A subsemigroup is a subset which forms a semigroup by itself under the product. An ideal is a subsemigroup such that every product of an element in it and an element outside it is in it. For instance an even integer times any integer is even. A quotient semigroup is a semigroup made up of equivalence classes of some equivalence relation under the same product.

Green's relations are defined from the ideals and enable any semigroup to be expressed as the union of equivalence classes. They to some degree express the relation between semigroups and groups.

Any set of binary relations generates a semigroup. Binary relations such as 'x associates with y' have been studied in particular groups of people by sociologists.

For any machine, such as a computer, there is a semigroup associated with the transitions from one internal state to another determined by inputs. This is a semigroup of functions under composition.

An abstract concept of finite-state machine has been defined involving a set of inputs, a set of outputs and a set of internal states. Such machines can accomplish various theoretical tasks such as adding numbers or recognizing a programming language. Programming languages themselves can be represented in terms of free semigroups.

A ***group*** is a semigroup having an identity element e such that $ex = xe = x$ for all x and an inverse x^{-1} of each element x such that $xx^{-1} = x^{-1}x = e$. Groups may be studied in terms of subgroups and quotient groups as semigroups can be. A homomorphism is a function f from one group to another such that $f(xy) = f(x)f(y)$. Finite (or finitely generated) commutative groups can be viewed as Cartesian products of cyclic groups (one generator groups), which are quotient groups of the integers.

A one-to-one onto function from a set to itself is called a ***permutation***. The permutations on any set form a group, and every group may be represented as a permutation group. The symmetries of a geometric set or mathematical model form a group, and use of symmetry often simplifies the analysis of a mathematical situation.

A system of distinct representatives consists of a selection of an object from each of n-sets $U_1, U_2, \ldots, U_n$ such that every selection is different from the rest. If $|U_1 \cup U_2 \cup \ldots \cup U_n| = n$ this must in effect be a permutation. The Hall-Koenig theorem gives a necessary and sufficient condition for a system of distinct representatives to exist. It is convenient to prove it by the important Ford-Fulkerson algorithm for flows on networks.

A group of permutations on a set is said to act on the set. The subsets of elements mapped into one another by the group are called ***orbits***. The size of each orbit equals the size of the group divided by the number of elements leaving a given number of the orbit fixed (isotropy group). This is the basis for many enumerational results where symmetry is involved. Polya's theory describes the number of classes of functions f from a set X acted on by a group G to another set S, where two functions differing by an action of G are equivalent. For instance the number of types of colorings of a cube by three colors can be computed using his theorem. Or the number of ways to arrange atoms in a given molecular diagram may be calculated.

In kinship systems of primitive tribe, there are sometimes a small number of clans such that the clan of children is determined from those of the parents in such a way as to prevent incest. These may be studied by means of groups.

A lattice is a partially ordered set in which any two elements have a least upper bound (as a union of sets) and a greatest lower bound (as an intersection). Subgroups of a group form a lattice.

2.1 SEMIGROUPS

A semigroup is about the simplest and most general mathematical structure of wide significance.

DEFINITION 2.1.1. A *semigroup* consists of a set S together with a binary operation $\circ$ mapping $S \times S$ to S and satisfying the associative law

$$x \circ (y \circ z) = (x \circ y) \circ z.$$

Thus a semigroup is a set provided with an associative binary operation. (It must be closed under the operation, that is, $a \circ b$ must always be within the set.)

EXAMPLE 2.1.1. Any of the following is a semigroup under either multiplication or addition: the positive integers, the integers, the positive rational numbers, the rational numbers, the positive real numbers, the real numbers, the set of $n \times n$ matrices over a field.

A semigroup will sometimes be written $(S, \circ)$ to indicate the operation.

Why should an operation be associative? For one thing, an associative operation on two elements defines unambiguously an operation on n elements, such as $x_1 + x_2 + \ldots + x_n$. For another associativity ties in with the idea of performing an operation on something like twisting a face of Rubik's cube or transforming raw materials into outputs. These are examples of transformations.

DEFINITION 2.1.2. A *transformation* on a set S is a function $f: S \rightarrow S$. The *composition* $f \circ g$ of two transformations is defined by $(f \circ g)(x) = g(f(x))$.

Many branches of mathematics use the reverse convention and call $(f \circ g)(x)$ what we call $(g \circ f)(x)$, so the student should be careful. However, our convention is the one that follows from the definitions of composition of binary relations in Definition 1.3.2 and is standard for semigroup theory.

The composition of two transformations means, apply the first transformation, then the second to its result. Associativity was proved in Proposition 1.3.1. Therefore any set of transformations closed under composition is a semigroup.

EXAMPLE 2.1.2. The following are semigroups, on any set S: all binary relations on S, all transformations on S, all onto transformations, all 1-to-1 onto transformations, all transformations sending a given subset T into itself, all transformations, having an image of at most k elements.

More generally all transformations on a set preserving some structure such as a partial order or a binary operation, form a semigroup.

Another class of operations giving rise to semigroups are least upper bound and greatest lower bound of two elements, provided these are unique.

EXAMPLE 2.1.3. For any subset of the real numbers, the operations $\sup\{x, y\}$ (the larger of x, y) and $\inf\{x, y\}$ (the smaller of x, y) are associative.

DEFINITION 2.1.3. A *lattice* is a partially ordered set S in which for any two elements a, b there exist unique elements $a \wedge b$, $a \vee b$ such that (1) $a \wedge b \leqslant a$, $a \wedge b \leqslant b$, (2) $a \vee b \geqslant a$, $a \vee b \geqslant b$, (3) if $c \geqslant a$, $c \geqslant b$ then $c \geqslant a \vee b$, (4) if $c \leqslant a$, $c \leqslant b$ then $c \leqslant a \wedge b$. The elements $a \vee b$, $a \wedge b$ are called the *least upper bound* and *greatest lower bound* of a, b, also the *join* and *meet* of a, b.

Not all posets are lattices, but many of the most important ones are.

EXAMPLE 2.1.4. The following are lattices: the real numbers, any linearly ordered set, the set of all subsets of a set, the set of all partitions of a set, the set of all lines and planes through the origin in E^3 together with $\{0\}$ and E^3 itself. Here E^3 is 3-dimensional space, and the partial order is inclusion in the last three examples.

In this book E^n will denote n-dimensional Euclidean space.

EXAMPLE 2.1.5. These partial orders are not lattices:

$$\{(1,1),(2,2)\}, \{(1,1),(2,2),(3,3),(3,1),(3,2)\}.$$

Every lattice is a special type of senigroup under $\vee$ (or $\wedge$) with several additional properties.

Commutativity: $a \vee b = b \vee a$

Idempotence: $a \vee a = a$

The last property is not true for semigroups obtained by addition or multiplication of integers or real numbers, and indicates something of the variety of semigroups that exist.

Finite lattices can in fact be described as finite idempotent commutative semigroups with identity.

A different type of semigroup is made up of sequences of symbols called ***words***. The product of two words is obtained by writing one word directly after another. For instance $(xyz)(xxzy) = xyzxxzy$. Associativity follows from $(w_1 w_2)w_3 = w_1 w_2 w_3 = w_1(w_2 w_3)$.

DEFINITION 2.1.4. The *free semigroup* generated by a nonempty set X is the set of all finite sequences from X (called words). The product of two sequences $x_1 x_2 \ldots x_r$ and $y_1 y_2 \ldots y_s$ is the sequence $x_1 x_2 \ldots x_r y_1 y_2 \ldots y_s$.

The importance of free semigroups comes from the fact that every semigroup is the image of a free semigroup under an onto function f such that $f(xy) = f(x)f(y)$.

EXAMPLE 2.1.6. The free semigroup generated by x, y has as its 5 elements $x, y, xy, yx, xx, yy, xyx, xxx, yyx, xyy, xxy, yyy, yxy, yxx$ and so on.

A free semigroup on one generator is commutative, but not free semigroups on more than one generator. All free semigroups on at least one free generator are infinite.

DEFINITION 2.1.5. A *subsemigroup* of a semigroup S is a subset $T \subset S$ such that if $x, y \in T$ then $xy \in T$.

EXAMPLE 2.1.7. The semigroup of positive integers is a subsemigroup of the semigroup of all integers.

Subsemigroups of a semigroup are subsets closed under the operation. Knowledge of the subsemigroups of a semigroup gives information about its structure.

DEFINITION 2.1.6. Two semigroups S, T are *isomorphic* if and only if there exists a 1-to-1 onto function $f\colon S \to T$ such that $f(x * y) = f(x) \circ f(y)$, where $*$ is the operation in S and $\circ$ is the operation in T.

EXAMPLE 2.1.8. The semigroup of real numbers under addition is isomorphic to the semigroup of positive real numbers under multiplication. The isomorphism is given by $f(x) = e^x$, and the property $f(x * y) = f(x) \circ f(y)$ corresponds to the basic properties of exponential functions $e^{x+y} = e^x e^y$. Here $*$ is $+$, $\circ$ is $\times$.

Two isomorphic semigroups are identical in mathematical structure. It can be considered that they differ only in the names of the elements: if x is renamed $f(x)$ then the products always agree. Thus if a semigroup S is proved isomorphic to a semigroup T it suffices to study T in order to determine the properties of S. For instance S is commutative or idempotent or finite or has an identity if and only if T has these respective properties. There is a 1-to-1 correspondence between subsemigroups of S and those of T.

Any given semigroup is isomorphic to a semigroup of transformations on a large enough set. Thus any abstract property true of all semigroups of transformations is true of every semigroup.

THEOREM 2.1.1. *Any semigroup S is isomorphic to a subsemigroup of a semigroup of transformations.*

Proof. Let 1 be an element disjoint from S. Make $S \cup \{1\}$ into a semigroup M in which S has its same product and 1 is an identity: $1x = x1 = x$. The associative law $(xy)z = x(yz)$ can be verified from by studying each case as to whether x, y, z equal 1 or are in S.

To each element $x \in M$ we associate a transformation f_x in M such that $f_x(a) = ax$. We have $f_{xy}(a) = axy = (ax)y = f_y(f_x(a)) = (f_x \circ f_y)a$. Therefore $f_{xy} = f_x \circ f_y$. Moreover this correspondence is 1-to-1 since if $f_x = f_y$ then $f_x(1) = x = f_y(1) = y$. This proves the theorem. $\square$

EXERCISES

Level 1

1. Give 3 additional examples of semigroups (what about complex numbers?).
2. Let f, g on $X = \{1, 2, 3\}$ be given by $f(1) = 2$, $f(2) = 3$, $f(3) = 3$, $g(1) = 3$, $g(2) = 1$, $g(3) = 3$. Calculate $f \circ g$ and $g \circ f$.
3. If h is given by $h(1) = 2$, $h(2) = 3$, $h(3) = 1$ verify the associative law for $(f \circ g) \circ h$. Here f and g are the same as in the above exercise.
4. What is $2 \vee 3$? $3 \wedge 2$?
5. Prove in any lattice that $a \wedge a = a$, $a \vee a = a$, $a \wedge b = b \wedge a$, $a \vee b = b \vee a$.

6. What is the product $(xyzz)(zyz)$? Give examples to illustrate the fact that free semigroups are associative but not commutative.

Level 2

1. Consider all transformations on real numbers of the form $ax + b$. Prove this is a semigroup (i.e. prove closure) and give a formula for the composition of $ax + b$ and $cx + d$.
2. What are $\wedge$, $\vee$ in the lattice of subsets of a set?
3. Prove that if $(S, \circ)$ is a semigroup then another semigroup can be defined as $(S, *)$ where $x * y = y \circ x$. This is called the ***opposite semigroup***.
4. Write down all transformations on the set $\{1, 2, 3\}$.
5. Compute the composition of all transformations on $\{1, 2, 3\}$ having at most two elements in their image.
6. Prove that transformations having image size $\leqslant k$ form a subsemigroup, and even more, an ideal: if f does so do $f \circ g$ and $g \circ f$ for all g.
7. Write out all words of length exactly 4 in the free semigroup on 2 letters x, y.

Level 3

1. How can the semigroup of $n \times n$ matrices over $\mathbf{R}$ be regarded as a semigroup of transformations?
2. Let S be a finite commutative, idempotent semigroup with an identity e such that $ex = xe = x$. Define a partial order such that $xy = x \vee y$. Prove it is a partial order.
3. Define $\wedge$ in the situation of the last example. Prove this gives a lattice.
4. How many transformations are there on a set of n elements?
5. Describe free semigroups on one generator. Give an isomorphism from such a free semigroup to the positive integers.
6. When will two elements of a free semigroup commute?

2.2 GENERATORS AND RELATIONS

A semigroup can be described by listing all its elements and the product of any elements.

EXAMPLE 2.2.1. The following is a semigroup:

	a	b	c	d
a	a	b	a	b
b	a	b	a	b
c	c	d	c	d
d	c	d	c	d

It is a special case of a ***rectangular band***, that is, a Cartesian product with multiplication $(a, b)(c, d) = (a, d)$.

In many cases a briefer description is possible. Instead of giving all elements, it is sufficient to give a set $G \subset S$ of elements such that every other element is a product of a sequence of elements of G. In this case G is called a ***set of generators*** for S.

DEFINITION 2.2.1. Let S be a semigroup and $G \subset S$. Then G is a *set of generators* for S if S is the set of all products $x_1 x_2 \ldots x_s$ such that $x_i \in G$.

For the positive integers under addition $\{1\}$ is a generating set since every positive integer has the form $1 + 1 + \ldots + 1$. A free semigroup on 2 generators illustrates the fact that a finite set of generators can generate an infinite semigroup.

In terms of transformations a generating set is a set from which all transformations of the semigroup can be produced. A generating set for the transformations on ***Rubik's cube*** has 6 elements, a quarter-turn clockwise for each face.

EXAMPLE 2.2.2. Any rearrangement of n objects can be obtained by successively interchanging pairs of adjacent objects. This implies that the permutations $(k\ k + 1)$ which interchange $k, k + 1$ and keep all other elements fixed, generate the set of all permutations on $\{1, 2, \ldots, n\}$.

To describe the multiplication in a semigroup, not only the generators are needed, but also which products of generators equal one another.

DEFINITION 2.2.2. A *relation* in a semigroup S is an equation $x_1 x_2 \ldots x_r = y_1 y_2 \ldots y_s$ for some $x_i, y_i \in S$.

EXAMPLE 2.2.3. In the additive semigroup of positive integers we have the relations $2 + 3 = 4 + 1, 6 + 9 = 5 + 10$.

A semigroup is essentially known if we know a set of generators and all relations among those generators. It is not necessary even to know all relations, just enough to imply all relations. Every element is a product of generators and the relations tell which products are distinct. The multiplication is described by $(x_1 x_2 \ldots x_r)(y_1 y_2 \ldots y_s) = x_1 x_2 \ldots x_r y_1 y_2 \ldots y_s$.

EXAMPLE 2.2.4. The semigroup of positive integers $2^n 3^m$, n, m $\geqslant 0$ under multiplication has generators 1, 2, 3. These are a set of defining relations:

$$1 \cdot 1 = 1, \quad 1 \cdot 2 = 2, \quad 2 \cdot 1 = 2, \quad 1 \cdot 3 = 3,$$

$$3 \cdot 1 = 3, \quad 2 \cdot 3 = 3 \cdot 2.$$

Congruences on a semigroup are equivalence relations such that the equivalence class of a product depends only on the equivalence classes of its factors. This means that the equivalence classes themselves form a semigroup, called the ***quotient semigroup***. If two elements are congruent and both are multiplied by the same element, the products are congruent.

DEFINITION 2.2.3. A *congruence* on a semigroup S is an equivalence relation E on S such that if $x\,E\,y$ and $z \in S$ then $xz\,E\,yz$ and $zx\,E\,zy$.

Let $x\,E\,y$ and $z\,E\,w$ then by definition of congruence $xz\,E\,yz\,E\,yw$. This proves that the equivalence class of a product depends only on that of its factors. Since associativity follows from $(\overline{x}\overline{y})\overline{z} = \overline{xy}\,\overline{z} = \overline{(xy)z} = \overline{x(yx)} = \overline{x}\,\overline{yz} = \overline{x}(\overline{y}\,\overline{z})$, we have a semigroup.

DEFINITION 2.2.4. The *quotient semigroup* defined by an equivalence relation E is the set of equivalence classes $\overline{x}$ with product $\overline{x}\,\overline{y} = \overline{xy}$.

EXAMPLE 2.2.5. A congruence on the positive integers is defined by $x\,E\,y$ if and only if $x = y$ or $x, y > z$. The quotient semigroup has three classes $\overline{1}, \overline{2}, \overline{3}$ and multiplication

	$\overline{1}$	$\overline{2}$	$\overline{3}$
$\overline{1}$	$\overline{2}$	$\overline{3}$	$\overline{3}$
$\overline{2}$	$\overline{3}$	$\overline{3}$	$\overline{3}$
$\overline{3}$	$\overline{3}$	$\overline{3}$	$\overline{3}$

The theory of generators and relations is connected with the theory of congruences on a free semigroup on the generating set. Each relation corresponds to two words in the free semigroup being equivalent.

We first state that any set of relations on a generating set gives a congruence on a free semigroup.

PROPOSITION 2.2.1. *Let F be the free semigroup generated by a set C. Let w_i, v_i for $i = 1$ to k be words, i.e. elements of F. Let R be the relation such that $(w, w') \in R$ if and only if w' can be obtained from w by a finite number of changes of the form $aw_ib \to av_ib$ or $av_ib \to aw_ib$, where $a, b \in F$ (as a special case we allow a or b to be a sequence of no elements). Then R is a congruence.*

Proof. Straightforward.

Call the congruence of this proposition $E(v_i = w_i)$.

DEFINITION 2.2.5. The semigroup with generating set C and defining relations $w_i = v_i$ is the *quotient semigroup associated with the congruence* $E(v_i = w_i)$ on the free semigroup generated by C.

EXAMPLE 2.2.6. The semigroup with generating set x and defining relation $x = x^4$, has 3 distinct elements x, x^2, x^3. (We have $x^4 = x$, $x^5 = x$, $x^4 = xx = x^2$, $x^6 = x^4x^2$, and so on.) Products are given by the table below.

	x	x^2	x^3
x	x^2	x^3	x
x^2	x^3	x	x^2
x^3	x	x^2	x^3

Namely $x^5 = x^4x = xx = x^2$, $x^6 = x^4x^2 = xx^2 = x^3$, and so on.

EXAMPLE 2.2.7. The semigroup with generators x, y and defining relation $xy = yx$ has as its distinct elements all products x^ny^n. Multiplication is given by $x^ny^mx^ry^s = x^{n+r}y^{m+s}$. Namely by induction one can prove that any power of x must commute with any power of y.

A ***homomorphism*** of semigroups is more general than an isomorphism, in that it need not be 1-to-1 or onto. Although a homomorphism, may not be 1-to-1, frequently a homomorphic image of a semigroup is simpler than the original and thus gives insight into its structure.

DEFINITION 2.2.6. A *homomorphism* of semigroups $f: S \to T$ is a function f such that $f(xy) = f(x)f(y)$.

EXAMPLE 2.2.8. There exists a homomorphism f of semigroups from the set of all positive integers under multiplication to the set of positive integers under addition such that $f(2^n(\text{odd number})) = n$. For instance $f(36) = f(4 \cdot 9) = f(2^2 \cdot 9) = 2$.

EXAMPLE 2.2.9. The *determinant* is a homomorphism of semigroups from $n \times n$ matrices under matrix multiplication to real (or complex) numbers under multiplication. This is true because

$$\det(AB) = \det(A)\det(B).$$

In some applications it is convenient to represent semigroups by generators and defining relations, or equivalently to find a set of generators and a set of relations which imply every relation. For any set of relations in a semigroup S, the next proposition shows there is a homomorphism of the semigroup defined by those relations, onto S. This will be an isomorphism if the set of relations is complete.

PROPOSITION 2.2.2. *Let C be a set of elements contained in a semigroup S and $w_i = v_i$ certain relations which hold in S, where w_i, v_i are products of elements of C. Let G be the semigroup with generating set C and defining relations $w_i = v_i$. Then there exists a unique homomorphism $f\colon G \to S$ such that $f(c) = c$ for all $c \in C$.*

Proof. Straightforward.

To check that a complete set of relations has been found, it may be necessary to determine the semigroup determined by a set of relations $v_i = w_i$ in certain generators. First deduce some consequences of these relations. Then use these to show that every word is equal to a word from a list of words S. (The set S should be such that no two words are equal under the relations. If not, delete one of the equal words.)

Compute products in S by taking the products of powers of words and reducing them to elements of S by means of the relations. If products are associative and the relations $v_i = w_i$ are satisfied then the semigroup determined by the given relation has been found.

EXAMPLE 2.2.10. Let a semigroup have generators x, y and defining relation $xy = yx^2$. Then by induction, $xy^s = y^s x^{2^s}$. By another induction $x^t y^s = y^s x^{2^s t}$. This means any word can be rearranged so that the y powers (if any) precede the x powers (if any). Thus any word can be written in the form $y^s x^t$ for $s, t \geqslant 0$.

Products are given by $y^s x^t y^u x^v = y^s y^u x^{2^u t} x^v = y^{s+u} x^{2^u t+v}$. It can be verified that these are associative. So the semigroup determined by this relation has been found.

For many elementary semigroups this procedure can readily be carried out. However, it can be very difficult. In fact, it has been proved that the problem of deciding whether or not two words are equal, in a semigroup specified by generators and relations, can be undecidable.

There is a different method for studying semigroups, without generators and relations, taken up in the next section.

EXERCISES

Level 1

1. A more general rectangular band can be definied on any nonempty Cartesian product set $S \times T$ by the multiplication $(i, j)(n, m) = (i, m)$. Prove this is associative.

2. Show that if S or T contains at least 2 elements, a rectangular band is not commutative.
3. Show these relations hold identically in rectangular bands: $x^2 = x$, $xwy = xzy$.
4. What are a set of generators for the positive integers under multiplication?
5. Show that in the semigroup S of constant transformations on a set S we have $xy = x$ for all x, y in S.
6. Consider a semigroup with one generator x and one relation $x^3 = x$. Prove that $x^{2n+1} = x$ and $x^{2n+2} = x^2$ for all n. Prove x^2 acts as an identity element e. Assuming $x \neq e$ write out the multiplication table

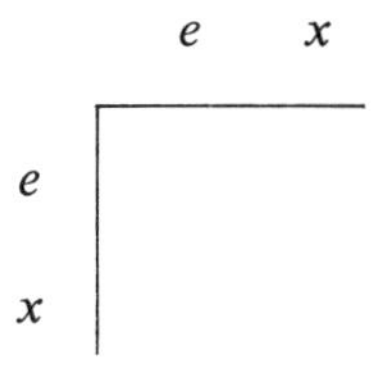

7. If e is an identity element in T and S is any semigroup show the constant mapping $f(x) = e$ is a homomorphism. Need e be an identity?

Level 2

1. Consider the transformations f_i, $i = 1, 2, \ldots, n-1$ on $\{1, 2, \ldots, n\}$ such that $f_i(i) = i + 1$, $f_i(i + 1) = i$ and $f_i(x) = x$ for other x. Prove $f_i \circ f_i = e$ where $e(x) = x$ for all x.
2. Prove $f_i \circ f_j = f_j \circ f_i$ whenever $j \neq i \pm 1$.
3. Prove $(f_i \circ f_{i+1})^2 = f_{i+1} \circ f_i$ for $i = 1, 2, \ldots, n-2$.
4. Describe all homomorphisms from the positive integers under addition to itself.
5. Write out the set of transformations from $\{1, 2\}$ into itself. Let x be the transposition interchanging 1, 2 and y the constant transformation $y(i) = 1$ for $i = 1, 2$. Prove the other two can be expressed as products of x, y.
6. Give as far as you can, a complete set of relations on x, y in the above exercise.
7. Describe the semigroup with generators x, y and defining relation $xy = yx$.

Level 3

1. What semigroup is defined by these relations on x, y: $xy = yx$, $xxy = x$, $yxy = y$?
2. Consider the semigroup defined by these relations: $xx = yyy$, $xxx = x$, $yyyy = y$, $xyx = yy$. How many elements does it have? What is $xx = yyy$?
3. Classify homomorphisms from a free semigroup into itself.

4. Show that for any set C any semigroup S and any function $f: C \to S$ there exists a unique homomorphism g from the free semigroup generated by C into S such that $f(c) = g(c)$ for $c \in C$. This is the '***universal property***' of free semigroups.
5. Show that if a semigroup has the property of the above exercise it is isomorphic to a free semigroup.
6. Consider the semigroup of all transformations $f(x) = ax + b$ where $a = \pm 1$ and b is an integer. Find a set of generators for this semigroup, and some relations among them.
7. Give some homomorphisms from the semigroup of $n \times n$ matrices under addition to the real numbers.

2.3 GREEN'S RELATIONS

The algebraic theory of semigroups is based on constructing semigroups out of simpler components. For instance one type of component might be a subset S which is a group (that is has an identity e such that $ex = xe = x$ for $x \in S$ and for all $x \in S$ there is an element $y \in S$ with $xy = yx = e$). Groups have their own structure, which we take up later, but from the viewpoint of semigroup theory they are irreducible and cannot be broken down further.

EXAMPLE 2.3.1. The semigroup of transformations contains the subset of 1-to-1 onto transformations, which is a group. It also has other subsets which are groups.

EXAMPLE 2.3.2. The multiplicative semigroup of rational numbers consists of 0 together with the multiplicative group of rational members.

Subsemigroups, subsets closed under multiplication, are significant in the structure of semigroups. Even more significant are ideals, subsets closed under multiplication even if one factor is outside the subset.

DEFINITION 2.3.1. A *left* (*right, two-sided*) *ideal* in a semigroup S is a non-empty set $K \subset S$ such that if $x \in K$, $y, z \in S$, $yx \in K$ ($xy \in K, xy, yx, yxz \in K$).

EXAMPLE 2.3.3. The set of $n \times n$ singular matrices over a field has the ideal consisting of all singular matrices, since if X has determinant 0, so does YXZ.

EXAMPLE 2.3.4. The set of positive integers under multiplication has an ideal I_m consisting of all multiples of m for a positive integer m. This is true because if x is a multiple of m so is yxz.

For noncommutative semigroups there can be left-ideals which are not right-ideals and vice versa. There are ideals associated with all elements x; however, these ideals may not be distinct and may be the entire semigroup, which is always a two-sided ideal.

DEFINITION 2.3.2. For subsets A, B of a semigroup, AB means $\{ab: a \in A, b \in B\}$, the set of products formed by an element of A times an element of B.

DEFINITION 2.3.3. The *left* (*right*) *ideal generated* by a subset A in a semigroup S is $SA \cup A$ ($AS \cup A$). The two-sided ideal generated by x is $\{x\} \cup \{xS\} \cup \{Sx\} \cup \{SxS\}$.

DEFINITION 2.3.4. A *principal ideal* is one generated by a single element.

EXAMPLE 2.3.5. The two-sided ideal generated by x in the additive semigroup of positive real numbers is all real numbers $y \geqslant x$.

EXAMPLE 2.3.6. In the multiplicative semigroup of positive real numbers 0 generates the ideal $\{0\}$.

For any semigroup S an element 1 can be added. If products in S are extended by $1a = a1 = a$, $a \in S$, and $1 \cdot 1 = 1$ this gives a new semigroup with identity, denoted S^1. Definitions can sometimes be stated more conveniently in terms of S^1 than S.

DEFINITION 2.3.5. Two elements of a semigroup are L-*equivalent* (R-*equivalent*, J-*equivalent*) if and only if they generate the same left (right, two-sided) ideal.

It is readily verified that the relations defined by this are equivalence relations in any semigroup.

EXAMPLE 2.3.7. In the semigroup of constant transformations any two elements are L-equivalent since $xy = x$ means Sy equals S. But no two elements x, y are R-equivalent since $yS = \{y\}$ and $xS = \{x\}$.

Both L-equivalence and R-equivalence imply J-equivalence but not conversely. Two more relations are built out of L, R.

DEFINITION 2.3.6. Two elements x, y are H-*equivalent* if and only if they are both R and L-equivalent. They are D-*equivalent* if and only if there exists z such that $x \mathsf{R} z$ and $z \mathsf{L} y$.

Thus H is the intersection relation $\mathsf{R} \cap \mathsf{L}$ and D is the composition $\mathsf{R} \circ \mathsf{L}$. In a commutative semigroup all 5 relations coincide. In a group any two elements are equivalent under each relation.

In a free semigroup no two elements are J-equivalent.

EXAMPLE 2.3.8. Two matrices over a field are J (D)-equivalent if and only if they have the same rank. They are R (L)-equivalent if and only if their row (column) spaces are identical. They are H-equivalent if and only if their rows span the same space and their columns span the same space.

The five relations mentioned in Definition 2.3.6 are known as ***Green's relations.*** These relations have a meaning in terms of solving equations like $xa = b$ or $ax = b$ in the semigroup.

THEOREM 2.3.1. *The Green's relations are equivalence relations. The relations* R, L, J *can alternatively be defined as follows, in a semigroup S:*

(1) a L $b \Leftrightarrow xa = b$ and $yb = a$ for some $x, y \in S^1$

(2) a R $b \Leftrightarrow ax = b$ and $by = a$ for some $x, y \in S^1$

(3) a J $b \Leftrightarrow zax = b$ and $wby = a$ for some $x, y \in S^1$

Proof. First (1) will be proved. Suppose a L b. Then $S^1a = S^1b$. So $a \in S^1b$ and $b \in S^1a$. So $a = yb$, $b = xa$ for some $x, y \in S^1$.

Suppose $a = yb$, $b = xa$. Then $a \in S^1b$ and $b \in S^1a$. Thus $S^1a \subset S^1(S^1b) = (S^1S^1)b = S^1b$. And $S^1b \subset S^1a$. So a L b. This proves (1).

The proofs of (2) and (3) are similar. □

COROLLARY 2.3.2. *The relation* D *is an equivalence relation.*

Proof. Suppose a D b. Then a R c and c L b for some c. So for some x, y, z, w, $ax = c$, $cy = a$, $zc = b$, $wb = c$ by (1), (2) of Theorem 2.3.1. To prove symmetry it will suffice to find d such that b R d and d L a. Let $d = za$. Then $wd = wza = wzcy = wby = cy = a$. So d L a. And $dx = zax = zc = b$. Also $by = zcy = za = d$. So d R b. This proves symmetry.

Note that D is the composition relation R $\circ$ L $=$ L $\circ$ R. But now D $\circ$ D $=$ R $\circ$ L $\circ$ R $\circ$ L $=$ R $\circ$ R $\circ$ L $\circ$ L $\subset$ R $\circ$ L $=$ D. This proves transitivity of D. It is reflexive since R, L are. □

Equivalence classes of these relations are called R, L, J, H, and D-*classes* respectively. The relation D can also be viewed as the smallest equivalence relation containing R, L (join of R, L).

The following tables gives some of these classes for several semigroups.

	R-classes	L-classes	J-classes = D-classes
$M_n(F)$	null space	image	rank
T_n	partition	image	rank
B_n	column space	row space	isomorphism type of row space

Here the semigroups $M_n(F)$, T_n, and $\mathcal{B}_n$ are, respectively, $n \times n$ matrices over the field **F**, transformations on the set $\{1, 2, \ldots, n\}$, and $n \times n$ Boolean matrices. The table gives a characteristic which determines the class of an element in each semigroup. Matrices are regarded as acting on row vectors. The partition of a function f is the set of equivalence classes of the relation $\{(x,y): f(x) = f(y)\}$.

THEOREM 2.3.3. *For any finite semigroup the relations* J *and* D *coincide.*

Proof. If a J b then $a = xby$, $b = zaw$ for some $x, y, z, w \in S^1$. We have the inclusion $aS^1 \supset awS^1$ and the epimorphism $awS^1 \to zawS^1$. So $|aS^1| \geqslant |awS^1| \geqslant |bS^1|$. Likewise $|bS^1| \geqslant |aS^1|$. So $|aS^1| = |awS^1| = |bS^1|$. So $aS^1 = awS^1$, since $aS^1 \supset awS^1$. So a R aw. Likewise $S^1aw \supset S^1zaw = S^1b$ and $|S^1b| \geqslant |S^1xb| \geqslant |S^1xby| \geqslant |S^1xbyw| = |S^1aw|$. So $S^1aw = S^1b$. So aw L b. Therefore a D b. □

Within a D-class the elements can be arranged in R and L-classes like bottles in a crate (this is called the *egg box picture*). Let $R_1, R_2, \ldots, R_n$ be the R-classes contained in D and $L_1, L_2, \ldots, L_m$ the L-classes contained in D. Then the H-classes are

$$\begin{array}{llll} R_1 \cap L_1 & R_1 \cap L_2 & \ldots & R_1 \cap L_m \\ R_2 \cap L_1 & R_2 \cap L_2 & \ldots & R_2 \cap L_m \\ \vdots & & & \\ R_n \cap L_1 & R_n \cap L_2 & \ldots & R_n \cap L_m \end{array}$$

These H-classes all have the same size, and there is an explict isomorphism between any two. For details on this and succeeding results, see Clifford and Preston (1964).

Regularity is a concept related to existence of certain weak inverses of an element.

DEFINITION 2.3.7. An element x of a semigroup S is *regular* if and only if there exists $y \in S$ such that $xyx = x$. (Here x is a little like an inverse of y.)

EXAMPLE 2.3.9. Any nonsingular matrix is regular, since we may take $y = x^{-1}$.

More generally, any element of a group is regular.

EXAMPLE 2.3.10. No element of the positive integers under addition, or of any free semigroup, is regular.

This is related to the fact that no inverse of an element in these semigroups (in a larger group) belongs in the semigroup.

A ***group inverse*** satisfies $xy = yx = e$, where e is an identity element. This implies $xyx = xe = x$, $yxy = ye = y$. If x, y satisfy the equations $xyx = x$, $yxy = y$ they are called ***generalized inverses***. The next theorem shows regularity implies existence of a generalized inverse. Generalized inverses are used in matrix theory for singular or non-square matrices, and statistics and other areas.

THEOREM 2.3.4. *Let S be a semigroup.*
(*1*) *$a \in S$ is regular if and only if a R e and a L f for certain idempotents e, f.*
(*2*) *If one element of a* D*-class is regular so is the entire* D*-class.*
(*3*) *An* L*-class* D *of a finite semigroup is regular if and only if* $\mathrm{DD} \cap \mathrm{D} \neq \emptyset$.
(*4*) *If a is regular, there exists $x \in S$ such that $a = axa$, $x = xax$.*

Proofs of all but (3) can be found directly in Clifford and Preston (1964) and (3) follows from their Lemma 2.17 by reasoning similar to the proof of Theorem 2.3.3.

Idempotents, H-classes, and subsets of semigroups which form groups are related. In a regular D-class some H-classes (those with idempotents in them) will be groups, and all other H-classes will be in 1–1 correspondence with groups.

EXERCISES

Level 1

1. Prove that in the additive group of integers any two elements are R-equivalent.
2. Prove that R = L = J in a commutative semigroup.
3. Let S be a semigroup with identity e with a partial order $>$ such that if $x > y$ then $ax > ay$. Show the elements $a > e$ form a subsemigroup T.
4. Let T be as in the above exercise. Prove that in T if x J y then $x = y$.
5. Prove that if x R y then x D y and x J y.
6. Show the subsets of a semigroup form a semigroup under the product AB.
7. Prove that every element of a group is regular.
8. What are the 5 relations in a rectangular band $S \times T$ with product $(r, s)(t, u) = (r, u)$?

Level 2

1. Let the rank of a transformation on $\{1, 2, \ldots, n\}$ be the number of elements in its image set. Prove rank $(fg) \leqslant$ rank (f) and rank $(fg) \leqslant$ rank (g).
2. Let A, B be subsets of $\{1, 2, \ldots, n\}$ having the same number of elements. Define a transformation f on $\{1, 2, \ldots, n\}$ such that $f(A) = B$.
3. Prove that two L-equivalent transformations f, g on $\{1, 2, \ldots, n\}$ must have the same image set.
4. Prove two R-equivalent transformations on $\{1, 2, \ldots, n\}$ have the same partition.
5. Prove two J-equivalent transformations have the same rank.
6. Show that the non-square matrix $[1 \quad 1]$ has a generalized inverse.
7. For transformation f on $\{1, 2, \ldots, n\}$, construct g as follows. For each $a \in$ Image (f) choose c_a such that $f(c_a) = a$. Let $g(a) = c_a$ for $a \in$ Image (f). For $x \notin$ Image (f) let g be any fixed element c_a. Prove g is a generalized inverse of f. Therefore the semigroup of transformations is regular.
8. What is the relation R in a lattice?

Level 3

1. Prove that if a semigroup has no left ideals except S and no right ideals except S then it is a group.
2. Prove that the relation $x \leqslant y$ if and only if $x = ayb$ for some a, b in S gives a quasi-order on S and gives a partial order on J-classes. Show this partial order is isomorphic to the set of two-sided principal ideals under inclusion.
3. Construct an infinite semigroup for which $\mathsf{J} \neq \mathsf{D}$ (use generators and relations).
4. A rectangular band of groups uses of nonempty sets S, T a group G, elements $g_{st} \in G$ for $s \in S$, $t \in T$ (sandwich matrix). The product on $S \times G \times T$ is given by $(r, x, t)\ (s, y, u) = (r, xg_{st}y, u)$. Prove this is associative. What are the 5 Green's relations?
5. Prove that two transformations with the same image set are L-equivalent.
6. Prove two transformations with the same partition are R-equivalent.
7. Prove two transformations of the same rank are D-equivalent.

2.4 BLOCKMODELS

Sociologists have studied groups of people in terms of various relationships that exist on the group: ***friendship, respect, influence, frequent contact, dislike.*** Each gives rise to a binary relation, as $\{(x, y): x \text{ is a friend of } y\}$. Each binary relation may then be represented by a Boolean matrix, whose (i, j)-entry is 1 if the relation exists between persons i and j, and whose (i, j)-entry is 0 if the relation does not exist. This gives a set of Boolean matrices for analysis.

One method of analysis is to try to find significant subsets of the group. An early method of R. D. Luce and A. D. Perry was to try to find ***cliques***: subsets of at least 3 members, and as large as possible, in which every person has the relation to every other. There are several hindrances: cliques are difficult to find by computer, and are usually approximate and not exact. R. D. Luce considered cliques in the square (or higher power) of a binary relation as a means of dealing with the second. However, some power of most binary relations is the Boolean matrix all of whose entries are 1.

Methods of finding subgroups of a binary relation which are approximate (in any sense) cliques are known as ***cluster analysis***. There are many methods: clusters can be built up from single individuals or the group may be repeatedly split in two so as to maximize some clustering index. Generalized components of a graph may be studied. Powers of a Boolean matrix may be taken.

In the method of blockmodels of H. C. White, S. A. Boorman, and R. L. Breiger, the individuals are to be grouped in blocks. Every individual in one block is to have the relation to every individual in another block, or else no individual in the former block is to have the relation to any individual in the second block. In practice, a few exceptions must be allowed.

This means we are searching for a ***congruence*** on the individuals with respect to the given relations. To group the individuals, a method, CONCOR, of R. L. Breiger, S. A. Boorman, and P. Arabie works well, although other clustering methods can be used. CONCOR repeats a step of passing from a Boolean matrix M to the Boolean matrix of correlations among rows of M, until this sequence of Boolean matrices converges.

To display a block pattern it is convenient to relabel the individuals so that those in the same block are adjacent. The relationships of the different blocks can be represented by a smaller Boolean matrix whose rows and columns correspond to blocks of individuals, that is, row i represents block i. To obtain it, consider each submatrix A_{ij} of the original Boolean matrix A, partitioned by blocks. Replance A_{ij} by a single 0 entry if A_{ij} is a zero matrix and replace it by a single 1 matrix if A_{ij} is not a zero matrix. This smaller Boolean matrix is called the *image*.

EXAMPLE 2.4.1. Griffith, Maier, and Miller (1976) studied a set of biomedical researchers. On of the relationships between researchers was that a pair of researchers had mutual contact with one another. Griffith, Maier, and Miller considered 107 men. White, Boorman, and Breiger investigated a random sample of 28 of these men.

When the group of men are suitably ordered, the matrix looks like this:

Mutual Contact

9		X	X	X	X				X						X												
26	X			X	X	X	X		X			X	X				X	X									
23	X								X																		
4	X	X							X							X											
1	X	X							X						X												
12		X					X	X					X					X									
7		X				X		X				X	X														
6						X	X						X	X													
2	X	X	X	X	X										X	X											
24																											
19																											
14		X					X																				
28		X				X	X	X																			
11								X																			
10	X				X				X																		
18				X					X																		
22		X																									
15		X				X																					
16																											
20																											
17																											
5																											
8																											
13																											
21																											
27																											
25																											
3																											

When the zero submatrices are replaced by 0 and the nonzero submatrices by 1, we obtain the following image:

$$\begin{bmatrix} 1 & 1 & 1 & 0 \\ 1 & 1 & 1 & 0 \\ 1 & 1 & 0 & 0 \\ 0 & 0 & 0 & 0 \end{bmatrix}$$

A hypothesis about the image matrix and a partitioning of the group which will produce this image is called a *blockmodel*. G. H. Heil developed a computer algorithm, which for any given blockmodel, finds all partitions of the set of people which yield the blockmodel as image.

Once a blockmodel is found it can sometimes be studied by partitioning it and producing images which are still simpler matrices.

Usually not one relation, but a number of binary relations are considered on the same group. All the resulting matrices should be partitioned the same way, so as to produce one image matrix for each binary relation.

The interrelationships of the resulting set of image matrices may be fairly complex. One technique, proposed by S. A. Boorman and H. C. White, is to study the subsemigroup of Boolean matrices generated by the image matrices. That is, we consider all matrix products of these matrices using Boolean operations on $\mathcal{B} = \{0, 1\}$.

Relations such as idempotence $A^2 = A$, commutativity $AB = BA$, and absorption $AB = A$ or $AB = B$ are of interest.

EXERCISES

Level 1

1. Describe the blocks and give the image for these Boolean matrices.

$$\text{(a)} \quad \begin{bmatrix} 1 & 1 & 0 & 0 \\ 1 & 1 & 0 & 0 \\ 0 & 0 & 1 & 1 \\ 0 & 0 & 1 & 1 \end{bmatrix}$$

$$\text{(b)} \quad \begin{bmatrix} 1 & 1 & 0 \\ 1 & 1 & 0 \\ 1 & 1 & 1 \end{bmatrix}$$

$$\text{(c)} \quad \begin{bmatrix} 1 & 1 & 0 & 1 \\ 1 & 1 & 0 & 1 \\ 1 & 1 & 1 & 0 \\ 0 & 0 & 0 & 1 \end{bmatrix}$$

2. Do the same for these after rearranging the individuals.

$$\text{(a)} \quad \begin{bmatrix} 1 & 0 & 1 \\ 0 & 0 & 0 \\ 1 & 0 & 1 \end{bmatrix}$$

(b) $\begin{bmatrix} 1 & 0 & 1 & 0 \\ 0 & 1 & 0 & 1 \\ 1 & 0 & 1 & 0 \\ 0 & 1 & 0 & 1 \end{bmatrix}$

Level 2

1. Explain the names of these blockmodels.

Deference: $\begin{bmatrix} 1 & 0 \\ 1 & 1 \end{bmatrix}$

Hierarchy: $\begin{bmatrix} 1 & 0 \\ 1 & 0 \end{bmatrix}$

Center-periphery: $\begin{bmatrix} 1 & 1 \\ 1 & 0 \end{bmatrix}$

Multiple caucus: $\begin{bmatrix} 1 & 0 \\ 0 & 1 \end{bmatrix}$

2. Give examples of relationships you would expect to lead to basically symmetric Boolean matrices. What about transitivity? Might it partly be true?
3. For which of these relations is $A^2 = A$?
4. Which pairs of these Boolean matrices commute?

Level 3

1. Suppose we consider Boolean matrices with a fixed partitioning, and square diagonal blocks. Let im (A) denote the *image matrix*. Prove im $(A + B) =$ im $(A) +$ im (B). The matrix addition is Boolean.
2. Prove that if all nonzero blocks in A, B have at least one nonzero row, im $(AB) =$ im (A) im (B). Here the matrix multiplication is Boolean.
3. Find all 2×2 Boolean matrices which equal their own correlation matrices.
4. Suppose a Boolean matrix is partitioned into 4 submatrices $A_{11}, A_{12}, A_{21}, A_{22}$ and each A_{ij} is a scalar multiple of the Boolean matrix having all entries 1. Suppose A equals its own correlation matrix. What possibilities are there for A?
5. Give an example of a binary relation on 5 elements such that a union of cliques contains a different clique.
6. Suggest an idea for a clustering method.

7. Write out the semigroup generated by these Boolean matrices:

$$\begin{bmatrix} 0 & 1 \\ 1 & 0 \end{bmatrix}, \begin{bmatrix} 1 & 0 \\ 0 & 0 \end{bmatrix}, \begin{bmatrix} 1 & 1 \\ 1 & 0 \end{bmatrix}$$

8. For two random Boolean matrices A, B and n large, the semigroup generated by A, B is likely to have three elements A, B, AB. Explain why.

2.5 FINITE STATE MACHINES

For an abstract model of a machine such as a calculator we consider basically three things: its ***inputs***, its ***outputs***, and its ***internal state***. Thus there will be a set X of inputs, a set Z of outputs, and a set S of internal states. Changes in internal states must also be taken into account. The new state depends on the previous state and the current input. Thus there is a function ν such that $\nu(s, x)$ is the next state if the current state is s and x is the input. Finally the output must be described. The output is a function of current state and (sometimes) input. It is described by a function δ such that $\delta(s, x)$ is the output from state s and input x.

DEFINITION 2.5.1. A *finite state machine* is a 5-tuple (S, X, Z, ν, δ) where S, X, Z are finite sets, ν is a function $S \times X$ to S, and δ is a function $S \times X$ to Z. Such a machine is called a *Mealy machine* after its inventor. A *Moore machine* is the same except that the output depends only on the internal state. Thus δ can be taken as a function $\delta : S \rightarrow Z$.

EXAMPLE 2.5.1. We can write out a machine to add two numbers in binary notation. At time i we put the digits a_i, b_i of the two numbers. Thus $X = \{0, 1\} \times \{0, 1\}$. The output is the ith digit of the output $\{0, 1\}$. So $Z = \{0, 1\}$. The internal state is the carry from previous digits, either 0 or 1. So $S = \{0, 1\}$. The output is the rightmost digit of the digits a_i, b_i, c where c is carry. So δ is given by

a_i	b_i	c	δ
0	0	0	0
0	0	1	1
0	1	0	1
0	1	1	0
1	0	0	1
1	0	1	0
1	1	0	0
1	1	1	1

And ν, the internal state, is the cary from $a_i + b_i + c$. So ν is 1 if and only if $a_i + b_i + c \geqslant 2$. Else ν is 0.

From this abstract machine an electronic switching circuit could be designed to do the addition. Machines can be designed to perform other logical or arithmetical tasks or in general do what machines or robots might do.

Finite state machines can also be used as models of general systems or of animal behaviour.

Semigroup theory, in particular the theory of transformations plays a role in the theory of finite state machines.

DEFINITION 2.5.2. The *semigroup of a machine* is the semigroup generated by the transformations $f_x(s)$ where x ranges over all inputs and $f_x(s) = \nu(s, x)$, together with the identity transformation on S.

EXAMPLE 2.5.2. The machine of Example 2.5.1 has semigroup $f_{00} = 0$, $f_{01} = f_{10}$ which is the identity, and $f_{11} = 1$. Thus the semigroup has 3 elements.

DEFINITION 2.5.3. A semigroup T having an element e such that $ex = xe = x$ for all $x \in T$ is called a *monoid*. Such an element e in any semigroup is called an *identity element*.

Thus the semigroup of a machine is a monoid, since the identity transformation is included.

EXAMPLE 2.5.3. The set of nonnegative integers under addition is a monoid but the semigroup of positive integers is not. It does not contain 0.

Machines can also be represented by graphs. A vertex is labelled for each internal state, and for each input x and state s an arrow labelled x is drawn from s to $\nu(s, x)$.

EXAMPLE 2.5.4. The graph of Example 2.5.3 is

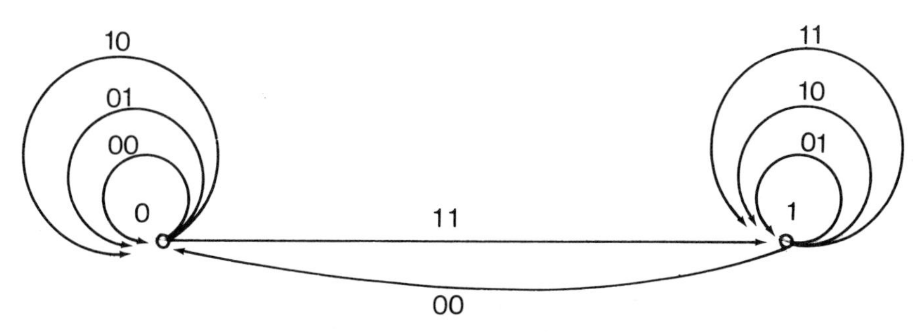

This amounts to graphing each transformation f_x.

These are some questions considered in the ***theory of automata*** (of which finite machines are the simplest kind):

1. *What is the machine having the fewest number of states which can accompl'sh a given task?*
2. *How can machines efficiently compute products in semigroups?*
3. *How can machines be decomposed into simpler machines?*
4. *What machine languages can machines recognize?*

The theory of automata more generally includes not only finite state machines but various types of machines with infinite memory: push-down automata and Turing machines. The latter can perform any calculation which can be done completely systematically. To show that a task cannot be performed by a Turing machine is considered equivalent to showing that no algorithm exists for it. This has been shown for a number of purely mathematical problems, for instance solving Diophantine equations (equations where the solutions are required to be whole numbers) in n variables.

We conclude this section by a result showing that no finite state machine can multiply two numbers of arbitrary length. This is contrary to the case for addition, where Example 2.5.1 constructs such a machine.

THEOREM 2.5.1. *Let M be a machine whose inputs are the sequence of digits a_i, b_i of two numbers a, b in binary notation and whose outputs are the successive digits of the product ab. Then if M can accept numbers of arbitrary length, M cannot have a finite number of states.*

Proof. Suppose M is a machine which can perform this task, having n states. Suppose we multiply the two numbers $a = 2^{n+1}$, $b = 2^{n+1}$. Let the initial state be s_0. The first $n+1$ inputs are $(0, 0)$. The next input is $(1, 1)$. Let s_1 be the state after this input. The next $n+1$ inputs are $(0, 0)$. The state after i of these is $(s_1)f_{(0,0)}{}^i$, $i = 0$ to n. But some two of the states $(s_1)f_{(0,0)}{}^i$ must be equal, since they lie in a set of n elements. Say $(s_1)f_{(0,0)}{}^i = (s_1)f_{(0,0)}{}^j$ where $i < j$. Then $(s_1)f_{(0,0)}{}^{n+1-(j-i)} = (s_1)f_{(0,0)}{}^{n+1}$, since the inputs and states are equal, the outputs at these times must be equal. But the output from the latter is the unique 1 digit of the answer, and the output from the former is 0. This proves the theorem. □

EXERCISES

In these exercises S is the set of states, X the set of inputs, Z the set of outputs, $\nu(s, x)$ is the next state from state s and input x and $\delta(s, x)$ is the output from state s and input x. For brevity ν and δ are written for $\nu(s, x)$ and $\delta(s, x)$.

Level 1

1. Consider a machine that will count to 100 (and then go back to 0). The output and the internal state are the number counted. How many internal states are needed?
2. The inputs will be either 0 or 1 as something is to be counted or not. Write S, X, Z.
3. The next state will be the same as the last if no input is received else it will be greater (or if the last is 100 the next is 0). Write the function $\nu(s, x)$ describing the next state.
4. Write the function δ giving the output.
5. Suppose the output were only to signify whether 100 is reached or not and after 100 is reached the state remains there. Write a machine for this.
6. Write a machine to test whether all digits of a binary number are zero.

Level 2

1. Consider a machine that will add three numbers in base 10, digit by digit. What possible internal states (carries) must be allowed for? Write S, X, Z.
2. Write ν, δ for a machine that adds two numbers in base 10.
3. Write a machine that adds three numbers in base 2.
4. Write a machine that multiplies any number in base 10 by 3.
5. Write a machine that divides any number in base 10 by 2. The digits must be put in left-to-right in contrast with previous examples.
6. Write a machine given two binary numbers a, b digit by digit to test whether $a = b$.
7. Write a machine given two numbers a, b in base 10 to decide whether $a = b$, $a < b$ or $a > b$.

Level 3

1. For any monoid S, any set X, and any mapping $h: X \to S$ construct a machine with semigroup S such that h is the mapping $h(x) = f_x$.
2. Write a machine that divides any number in base 10 by 11.
3. Can a finite state machine take the square of a number of arbitrary length input digit by digit right-to-left?
4. What tasks could be accomplished by a machine which has no memory, i.e. only one internal state? which remembers only the last input?
5. Show that some finite state machine can handle any task involving only a bounded number of inputs.

2.6 RECOGNITION OF MACHINE LANGUAGES BY FINITE STATE MACHINES

In recent years a very abstract notion of languages (mostly used for computer languages) has been developed. Here we consider only the purely formal aspects (***syntax***) of these languages and not their meaning (***semantics***).

We start with a set X of basic units (words) and consider sequences of words.

DEFINITION 2.6.1. If X is a set, X^* is the sequence of finite sequences from X together with the empty sequence e.

EXAMPLE 2.6.1. If X is $\{a, b\}$, X^* is $\{e, a, b, aa, ab, ba, bb, aaa, \ldots\}$.

DEFINITION 2.6.2. A *phrase-structure grammar* is a quadruple (N, T, P, σ) such that $T \neq \emptyset$, $N \cap T = \emptyset$, $P = ((N \cup T)^* \backslash T^*) \times (N \cup T)^*$, $\sigma \in N$, and N, T, P are finite sets. The set T is a set of words in the language, as sentence. The set N is a set of valid grammatical forms, which involve abstract concepts like 'variable', 'expression'. The set P is the set of ways we can substitute into a valid grammatical form to obtain another grammatical form. The element σ is called the *starting symbol*. The sets N, T, P are abbreviations for ***nonterminals, terminals, productions.*** That is for a nonterminal either a terminal (word) or nonterminal (more detailed grammatical form) can be substituted as allowed by P.

EXAMPLE 2.6.2. Suppose we want to obtain the statement $x = y + z + w$. Let T be $\{x, =, y, z, w\}$. Let $N = \{\sigma, \text{expression}, \text{variable}\}$. We start with σ and obtain successively

1. σ
2. Variable = Expression
3. Variable = Expression + Expression
4. Variable = Expression + Expression + Expression
5. Variable = Expression + Expression + w
6. Variable = Expression + $z + w$
7. Variable = $y + z + w$
8. $x = y + z + w$

The set of productions involved is (σ, variable = expression), (expression, expression + expression), (variable, x), (expression, y), (expression, z), (expression, w). Other sets N, P could also be used.

Steps (1)-(8) are called a *derivation*.

DEFINITION 2.6.3. If $x, y \in (N \cup T)^*$ then y is directly derived from x if $x = azb$ and $y = awb$ for some $(z, w) \in P$ and $a, b \in (N \cup T)^*$. An *indirect derivation* is a sequence of direct derivations.

That is, a direct derivation is a substitution of w in place of a nonterminal z where (z, w) is a valid production.

EXAMPLE 2.6.3. Going from 'variable = expression' to 'variable = expression + expression' is a direct derivation.

DEFINITION 2.6.4. $L(G)$ is the set of elements of T^* which can be derived from σ.

EXAMPLE 2.6.4. For the case above $L(G)$ includes $x = y$, $x = z$, $x = w$, $x = y + w$, $x = y + z$, $x = y + y$, $x = y + z + w + y$, and other equations.

DEFINITION 2.6.5. A grammar is *context-free* if and only if for all $(a, b) \in P$ we have $a \in N$, and $b \neq e$.

EXAMPLE 2.6.5. If x and y (but not xy) were members of N and $(xy, z) \in P$ then the grammar would not be context-free.

Being context-free means that what is substituted depends only on a single grammatical element, not surrounding elements.

DEFINITION 2.6.6. A grammar is *regular* if and only if for all $(a, b) \in P$ we have $a \in N$, $b = tN$ where $t \in T$ and $n \in N$ or $n = e$.

Being regular means that the productions are filled in with terminals one per step, going left to right and that at each stage there is a string of terminals followed by a single nonterminal.

EXAMPLE 2.6.6. Let $P = \{(\sigma, w), (w, xw), (w, yw), (w, x)\}$ and $T = \{x, y\}$. Then $L(G)$ is all sequences in x, y ending in x. This is a regular grammar.

DEFINITION 2.6.7. A machine *accepts* $L(G)$ if and only if the machine can tell whether or not any given sequence of inputs is in $L(G)$.

EXAMPLE 2.6.7. The following machine accepts the last language. The internal states are $\{0, 1\}$. Outputs are {Yes, No} (meaning all symbols so far do or do not constitute a valid expression).

$\nu(s, x)$

s	x	
0	x	0
0	y	0
0	other	0
1	x	1
1	y	1
1	other	0

$\delta(s, x)$

s	x	
1	x	Yes
other		No

We assume the initial state is 1.

THEOREM 2.6.1. *A language is acceptable by some finite state machine if and only if it is regular.*

Proof. Let $L(G)$ be accepted by a finite state machine. Outputs will be 'Yes, No' as in Example 2.6.7, meaning 'all symbols so far are a valid expression' or 'all symbols so far are not a valid expression'. The set T will be as in $L(G)$ but we will construct a new grammar based on the machine.

Let the set of nonterminals be σ together with the rest of internal states of the machine. Let the set of productions be the set of pairs (n_1, xn_2) such that if the machine is in state n_1 and x is input then state n_2 results, together with the set of (n_1, x) such that in state n_1 if x is the next input the machine declares the sentence to be valid.

Let $x_1 x_2 \dots x_k$ be an expression the machine recognizes as valid by a succession of internal states $n_1, n_2, \dots, n_k$ before these inputs. Then we have a derivation $x_1 x_2 \dots x_r n_r \rightarrow x_1 x_2 \dots x_{r+1} n_{r+1}$ in the language.

Conversely if $x_1 x_2 \dots x_n$ is derivable in the language and the sequence at time r is $x_1 x_2 \dots x_r n_r$ then n_r will be the state of the machine at time r for these inputs and the machine will recognize the sentence when the last input x_k is put in. This proves that if $L(G)$ is acceptable it arises from a regular grammar.

Conversely let $L(G)$ arise from a regular grammar. Let outputs for a machine be 'Yes, No' as above, and inputs be symbols of T. Let internal states be in 1-1 correspondence with all sets of nonterminals. Let the initial state be 0. If the state at time r is a set U of nonterminals let the state at time $r + 1$ be the set of all nonterminals z such that $(u, xz) \in P$ for some $u \in U$. Let the output be 'Yes' if and only if for some $u \in U$, $(u, x) \in P$. Then if we have any derivation $x_1 x_2 \dots x_r n_r \rightarrow x_1 \dots x_r x_{r+1} n_{r+1}$ then n_r will belong to U at time r and when x_k is put in the machine will print out 'Yes'. Conversely if the machine prints 'Yes' then there must have been a sequence of elements of the sets U making up a valid derivation. □

EXERCISES

In these exercises N is the set of nonterminals, T the set of terminals, P the set of productions, σ the starting symbols, $L(G)$ the language generated by grammar G, $\delta(s, x)$ the output from state s and input x and $\nu(s, x)$ the next state from state s and input x.

Level 1

1. In the grammar $T = \{+, 1\}$, $N = \{\sigma\}$, $P = (\sigma, 1)$, $(\sigma, \sigma + \sigma)\}$ derive the expression $1 + 1 + 1$.
2. What are the elements of $L(G)$?

3. Can $1 + 1 + 1$ be derived in two ways?
4. Is this grammar context-free? Is it regular?
5. In the grammar $T = \{x, (\ ,\ )\}, N = \sigma, P = \{(\sigma, x), (\sigma, \sigma\sigma), (\sigma, (\sigma))\}$ derive $(x)(x)$.
6. What is $L(G)$?

Level 2

1. What is $L(G)$ for this grammar $\{(\sigma, 1), (\sigma, 1 + \sigma)\}$ where N, T are as in Exercise 1 of Level 1.
2. Is the above grammar regular?
3. Do you think a regular grammar exists for $L(G)$ as in Exercise 5 of Level 1?
4. Find a regular grammar which generates $L(G) = \{$all sequences in x of length $(am + b)\}$ where a, b are fixed. Here $T = \{x\}$, and it is only necessary to find P, N.
5. Give a grammar to obtain a union $L(G_1) \cup L(G_2)$.
6. Let S be any finite subset of T^* for fixed T. Does there exist a regular grammar yielding S as $L(G)$?

Level 3

1. Two machines (S, X, Z, ν, δ), (T, X, Z, μ, ω) have the same behavior if there exists a binary relation R from S to T such that R and R^{T} are onto and if $(s, t) \in R$ and the machines are initially in states s, t, then for any sequence of inputs the machines give the same sequence of outputs. Show this is an equivalence relation.
2. Two states s_1, s_2 of machine (S, X, Z, ν, δ) are equivalent if and only if for any sequence $x_1 x_2 \dots x_n$ of inputs $\delta((s_1) f_{x_1} f_{x_2} \cdots f_{x_{n-1}}, x_n) = \delta((s_2) f_{x_1} f_{x_2} \cdots f_{x_{n-1}}, x_n)$. Show this is an equivalence relation.
3. For any machine M show there exists a machine $\overline{M}$ having the same behavior whose states are equivalence classes of states of M.
4. Show that if M and N have the same behavior there exists a homomorphism of machines from $\overline{M}$ to N which is 1-1 on states of M. Here a homomorphism of machines from (S, X, Z, ν, δ) to (T, X, Z, μ, ω) is a function $f: S \to T$ such that for all $s \in S, x \in S$, $\delta(s, x) = \omega(f(s), x)$ and $f(\nu(s, x)) = \mu(f(s), x)$.
5. Tell why the preceding exercises show that for any machine M, $\overline{M}$ is a machine having the same behavior with the least number of states.

2.7 GROUPS

A ***group*** is a semigroup in which multiplication by any element is a 1-to-1 onto function.

DEFINITION 2.7.1. A *group* consists of a set G and an operation $\circ$; $G \times G \to G$ such that

(1) $(x \circ y) \circ z = x \circ (y \circ z)$
(2) there exists $e \in G$ such that for all $x \in G$, $x \circ e = e \circ x = x$, where e is called an *identity element*
(3) for all $x \in G$ there exists $x^{-1} \in G$ such that $x \circ x^{-1} = x^{-1} \circ x = e$, where x^{-1} is called the *inverse* of x.

The presence of inverses x^{-1} makes group theory very different from semigroup theory. Every multiplication is reversible and cancellation holds.

EXAMPLE 2.7.1. Under addition, the following are groups: the integers **Z**, the rational numbers **Q**, rational numbers $\mathbf{Z}(\frac{1}{\mathbf{p}})$ whose denominator is a power of p, the real numbers **R**, the complex numbers **C**, $n \times m$ matrices for any n, m.

Positive integers, rationals, or reals do not form a group, lacking the identity 0 and the negative elements which are inverses.

EXAMPLE 2.7.2. Under multiplication these are groups: rationals r/s such that r, s have among their divisors a fixed set S of primes, nonzero rational numbers, nonzero real numbers, nonzero complex numbers, complex numbers of absolute value 1, nonsingular $n \times n$ matrices over **C**.

DEFINITION 2.7.2. A 1-to-1 transformation from a set to itself is called a *permutation*.

EXAMPLE 2.7.3. The permutations on any set S form a group. Associativity follows from their being a subsemigroup of the semigroup of transformations. The identity is given by $e(x) = x$ for all $x \in S$. The inverse is defined by $f^{-1}(y) = x$ whenever $f(x) = y$. A composition of 1-to-1 onto transformations is again a permutation.

For groups there is a result similar to the result about semigroups being isomorphic to semigroups of transformations.

PROPOSITION 2.7.1. *Any group G is isomorphic to a set of permutations on the set G.*

Proof. Essentially the same as the proof of Theorem 2.1.1.

DEFINITION 2.7.3. The group S_n of permutations on the set $\{1, 2, \ldots, n\}$ is called the *symmetric group of degree n* (or *symmetric group on n letters*).

The degree n of S_n should be distinguished from the order of a group G which is $|G|$. A symmetric group of degree n has order $n!$.

EXAMPLE 2.7.4. The symmetric group on $\{1, 2\}$ has two elements e, x where $x(1) = 2$, $x(2) = 1$ and multiplication $e^2 = e$, $ex = xe = x$, $x^2 = e$.

The symmetric groups will be studied in more detail later.

DEFINITION 2.7.4. A set C contained in a group G is a *set of generators* for G if and only if every element of G can be expressed as a product $x_1 \circ x_2 \circ \ldots \circ x_n$ where $x_i \in C$ or $x_i^{-1} \in C$.

EXAMPLE 2.7.5. The element 1 generates the integers as a group (but in the semigroup sense it would generate only the positive integers).

An equivalent statement is that no group which is a proper subset of G contains C.

DEFINITION 2.7.5. A *subgroup* of a group G is a subset which forms a group under the same operation.

This is equivalent to saying that a subgroup is a subset closed under products and inverses.

EXAMPLE 2.7.6. The additive integers are a subgroup of the additive rationals which are a subgroup of the additive real numbers which are a subgroup of the additive complex numbers.

The propositions on generators and relations for semigroups also go through for groups except that we must allow inverses. A ***free group*** is made up of words in the given elements and their inverses. When a product is taken any adjacent pair formed by an element and its inverse must be cancelled.

EXAMPLE 2.7.7. The free group on generators x, y has these words of length at most 2: $e, x, y, x^{-1}, y^{-1}, x^2, y^2, x^{-2}, y^{-2}, xy, x^{-1}y^{-1}, xy^{-1}, x^{-1}y, yx, y^{-1}x^{-1}$, yx^{-1}, $y^{-1}x$. The product of x^2yx^{-1} and xy^{-2} is $x^2yx^{-1}xy^{-2} = x^2yy^{-2} = x^2y^{-1}$ after cancelling $x^{-1}x$ and yy^{-1}.

The simplest nontrivial groups are those having one generator. For instance, they are always commutative.

DEFINITION 2.7.6. A group is *cyclic* if it has a single generator.

EXAMPLE 2.7.8. The additive group $\mathbf{Z}$ of integers is cyclic, with generator 1.

EXAMPLE 2.7.9. For any positive integer m there is a group $\mathbf{Z_m}$ whose elements are $0, 1, 2, \ldots, m-1$. Let the operation be $x * y =$ remainder if $x + y$ is divided by m. That is, it is the unique integer $0 \leqslant z \leqslant m$ such that m divides $x + y - z$. From this it can be shown the operation is associative. The inverse of a is $m - a$ if $a \neq 0$ and 0 is its own inverse.

The addition table for $\mathbf{Z}_3$ is

	0	1	2
0	0	1	2
1	1	2	0
2	2	0	1

EXAMPLE 2.7.10. Any 1-element group has this structure: $\{e\}$ where $e * e = e$. We will call this $\mathbf{Z}_1$.

THEOREM 2.7.2. *Any cyclic group G is isomorphic to* $\mathbf{Z}$ *or* $\mathbf{Z}_\mathbf{m}$.

Proof. Let x be a generator. Then all elements by definition can be expressed as positive or negative powers x^k of x. And the law of exponents $x^r x^s = x^{r+s}$ holds. Suppose there is no positive integer m such that $x^m = e$. Then $x^r \neq x^s$ for $r > s$ else $x^r x^{-s} = x^s x^{-s}$, $x^{r-s} = e$. So there is a 1-to-1 homomorphism from G to the integers defined by $f(x^n) = n$.

If there exists a positive integer m such that $x^m = e$, let m be the smallest such number. By induction we can readily prove that all powers of x equal one of $x^0, x^1, \ldots, x^{m-1}$ for instance $x^m = x^0$, $x^{m+1} = xx^m = xx^0 = x^1$, and so on. The powers $x^0, x^1, \ldots, x^{m-1}$ are distinct since if $x^r = x^s$, $m > r > s > 0$ then $x^r x^{-s} = x^s x^{-s} = e$, $x^{r-s} = e$. This contradicts the fact that m was the smallest positive integer such that $x^m = e$. Thus we have a 1-to-1 onto map from G to $\mathbf{Z}_\mathbf{m}$ defined by $f(x^i) = i$ for $i = 0$ to $m - 1$. It can readily be checked that this is a homomorphism, for instance by induction. This proves the theorem. □

Another way of constructing groups is by taking products.

DEFINITION 2.7.7. The Cartesian product $A_1 \times A_2 \times \ldots \times A_n$ of groups $A_1, A_2, \ldots, A_n$ is the group with set the Cartesian product set and operation $(g_1, g_2, \ldots, g_n)(h_1, h_2, \ldots, h_n) = (g_1 \circ h_1, g_2 \circ h_2, \ldots, g_n \circ h_n)$.

EXAMPLE 2.7.11. The n-fold Cartesian product of the real numbers is isomorphic to the set of n-dimensional vectors $(x_1, x_2, \ldots, x_n)$ with operation $(x_1, x_2, \ldots, x_n) + (y_1, y_2, \ldots, y_n) = (x_1 + y_1, x_2 + y_2, \ldots, x_n + y_n)$.

One use of Cartesian products is in studying the structure of commutative groups. It can be proved, though we will not prove it, that any finite commutative group is isomorphic to a product of cyclic groups $\mathbf{Z}_\mathbf{m}$ where the m's are powers of primes.

Commutative groups are frequently called *Abelian*.

EXERCISES

Level 1

1. Prove that any group of order 2 is isomorphic to $\mathbf{Z}_2$.
2. Prove that these cancellation properties hold in groups: if $xy = xz$ then $y = z$, if $yx = zx$ then $y = z$ (multiply by the x^{-1} on left or right).
3. Write out the addition table of $\mathbf{Z}_4$.
4. Write out the addition table of $\mathbf{Z}_2 \times \mathbf{Z}_2$.
5. Prove $\mathbf{Z}_4$ is not isomorphic to $\mathbf{Z}_2 \times \mathbf{Z}_2$.
6. Prove the symmetric group on $\{1, 2, 3\}$ is not commutative. Let f interchange 1, 2 and g interchange 2, 3. Prove $(f \circ g)(1) \neq (g \circ f)(1)$.
7. Prove the identity element of a group is unique.
8. Prove inverses are unique.

Level 2

1. Write out the multiplication tables of $\mathbf{Z}_2, \mathbf{Z}_3, \mathbf{Z}_2 \times \mathbf{Z}_3$.
2. Show $\mathbf{Z}_2 \times \mathbf{Z}_3$ is cyclic. Show $\mathbf{Z}_2 \times \mathbf{Z}_3$ is isomorphic to $\mathbf{Z}_6$.
3. Can you suggest a generalization of the above exercise about products of certain cyclic groups being cyclic?
4. Find all subgroups of $\mathbf{Z}_6$ (they are cyclic).
5. Prove the inverse of xy is $y^{-1}x^{-1}$. Describe the inverse any word $x_1x_2 \ldots, x_n$ in a free group.
6. Let $f(v_1, \ldots, v_n)$ be any function from n-tuples of vectors to the real numbers. Prove the set of invertible matrices A such that $f(v_1, v_2, \ldots, v_n) = f(Av_1, Av_2, \ldots, Av_n)$ is a group. For instance the set of all matrices A such that $|v|^2 = |Av|^2$ is a group called the *orthogonal group*. Its members preserve distances $|x - y|^2$ and are made up of totations and reflections in n-dimensional space.

Level 3

1. Consider the group of transformations of a square into itself including all rotations by multiples of 90° and all reflections through a line inclined at a multiple of 45° through the origin. Write out its multiplication table. This is the dihedral group of order 8.
2. Show there exists an onto homomorphism $f: \mathbf{Z}_m \to \mathbf{Z}_n$ if and only if n divides m. Show the set of x such that $f(x) = e$ is a subgroup.
3. Show that if n divides m, $\mathbf{Z}_m$ has a subgroup of order n.
4. Prove that any finite group is isomorphic to a group of matrices. Use Proposal 2.4.1 and prove that the symmetric group on $\{1, 2, \ldots, n\}$ is isomorphic to a set of $(0, 1)$-matrices with exactly one 1 in each row and column.
5. Suppose there exist homomorphisms $f_1 : G \to G_1$ and $f_2 : G \to G_1$. Construct a homomorphism $f: G \to G_1 \times G_2$. Use this to prove $\mathbf{Z}_{nm}$ is

isomorphic to $\mathbf{Z_n} \times \mathbf{Z_m}$ if n, m are relatively prime. (Since they have the same order it suffices to prove a homomorphism is 1-to-1. Let $f(x) = f(y)$. Then $f(xy^{-1}) = e$. Prove $xy^{-1} = e$.)

6. Prove that for any group G the onto homomorphisms $G \to G$ form a group. Such homomorphisms are called ***automorphisms.***
7. Let G, H be groups, and f a homomorphism from G into the automorphism group of H. Thus f_g is a homomorphism of H and $f_g \circ f_h = f_{hg}$. Define a product on $G \times H$ by $(g_1, h_1) \circ (g_2, h_2) = (g_1 g_2, f_{g_2}(h_1) h_2)$. Prove this product is associative. It gives a group called the ***semidirect product.*** Express the group of functions $f(x) = ax + b$, $a \neq 0$, a, b real numbers as a semidirect product.

2.8 SUBGROUPS

It has already been mentioned that a subgroup of a group is a nonempty subset closed under multiplication and inversion.

DEFINITION 2.8.1. For a subset C of a group G, the *subgroup* generated by C is the set of all products $x_1 x_2 \ldots x_n$ where $x_i \in C$ or $x_i^{-1} \in C$ for $i = 1$ to n.

It can readily be shown that this set is closed under products and inverses, and is therefore a subgroup. It is contained in every other subgroup containing C (since $x_1 x_2 \ldots x_n$ must be) and is thus the smallest subgroup containing C.

This means that for every subset C we obtain a subgroup generated by C, although many of these will be identical. For an element x we get a cyclic subgroup generated by x, consisting of all powers x^i.

EXAMPLE 2.8.1. A set of positive rational numbers $\{x_1, x_2, \ldots, x_k\}$ generates the multiplicative subgroup of rational numbers of the form $x_1^{n_1} x_2^{n_2} \ldots x_k^{n_k}$.

For any subgroup H we can divide a group into equivalence classes called ***cosets***. The cosets are sets which are 'parallel' to the subgroup in the sense $y = x + 1$ is parallel to $y = x$. All the lines $y = x + c$ partition the plane.

As before, AB means the set of all products $\{ab : a \in A, b \in B\}$. We have $(AB)C = A(BC) = \{abc : a \in A, b \in B, c \in C\}$.

DEFINITION 2.8.2. Let H be a subgroup of G. A *left* (*right*) *coset* of H is a set of the form xH (Hx) for some x in G. A *normal subgroup* H is one such that $xH = Hx$ for all x in G.

EXAMPLE 2.8.2. For a positive integer m, let H be the subgroup of $\mathbf{Z}$ of all integers divisible by m. Then the coset $1 + H$ consists of all integers having the form $1 + km$ for some integer k. The coset $0 + H$ equals H. Since $\mathbf{Z}$ is commutative, H is normal.

For brevity, let $a | b$ denote a divides b.

THEOREM 2.8.1. *For any subgroup H of a group G, G is the disjoint union of the left cosets of H. Each left coset of H has the same cardinality as H. If G is finite then $|H|\,|\,|G|$ and $|G|/|H|$ is the number of cosets. If H is a normal subgroup of G, then the left cosets of H form a subgroup under the operation $(xH)(yH) = xyH$.*

Proof. The left cosets of G are the equivalence classes of the equivalence relations $\{(x, y): x = yh$ for some $h \in H\}$. The first assertion follows from this. The mapping $H \to xH$ which sends h to xh is 1-to-1 and onto. This implies the second statement. The first two statements imply $|G|$ is $|H|$ times the number of cosets. If H is normal then $xHyH = x(yH)H = xyHH$. But $HH = H$ since H is a subgroup. So the operation is well defined. The proof of associativity is $(xHyH)zH = (xyH)zH = (xy)zH = x(yz)H = xHyzH = xH(yHzH)$. And $x^{-1}H = (xH)^{-1}$ and $eH = H$ is an identity element. This proves the last statement. □

EXAMPLE 2.8.3. The symmetric group of degree 3 has multiplication table:

	e	a	b	x	y	z
e	e	a	b	x	y	z
a	a	b	e	y	z	x
b	b	e	a	z	x	y
x	x	z	y	e	b	a
y	y	x	z	a	e	b
z	z	y	x	b	a	e

The set $H = \{e, a, b\}$ is a subgroup. There are two different left cosets of H, $eH = \{e, a, b\}$ and $xH = \{xe, xa, xb\} = \{x, y, z\}$. All other cosets equal one of these two. These two cosets form a partition of G. The subgroup H is normal. As can be seen from the distribution of symbols e, a, b, x, y, z in the table, multiplication of cosets is described by

	eH	xH
eH	eH	xH
xH	xH	eH

DEFINITION 2.8.3. If H is normal, the group of left cosets of H is called the *quotient group* G/H.

EXAMPLE 2.8.4. G/G is the one element group.

For a normal subgroup N, the ***order*** of G is given by $|N|\,|G/N|$ since there are $|G/N|$ cosets. Many other properties of G can be studied in terms of the simpler groups N, G/N if a proper nontrivial normal subgroup exists. In fact all groups G such that $G/N = H$ can be classified by means of extension theory. One such group is the *direct product* $N \times G/N$.

Groups having no normal subgroups can be studied in terms of other types of subgroups. If p^m is the highest power of a prime p such that p^m divides $|G|$ for $m \geqslant 1$, a subgroup of G of order p^m is called a ***Sylow p subgroup***. For any p dividing $|G|$ there exists a Sylow p subgroup of G and any two Sylow p subgroups for the same p are isomorphic, in fact for such subgroups H_1, H_2 there exists $x \in G$ with $xH_1x^{-1} = H_2$.

Groups of order p^m for prime p have a special structure. For instance, there always exists a nontrivial center.

DEFINITION 2.8.4. The *center* of a group G is $\{z \in G : xz = zx \text{ for all } x \in G\}$.

EXAMPLE 2.8.5. The center of a commutative group is the entire group.

EXAMPLE 2.8.6. The center of a symmetric group of order > 2 is the identity element only.

EXAMPLE 2.8.7. The center of the group of $n \times n$ nonsingular matrices consists of all scalar multiples aI of the identity matrix.

Thus the center is the set of elements commuting with every element of G.

DEFINITION 2.8.5. A group is *simple* if it is not $\{e\}$ and has no normal subgroups except $\{e\}$ and the entire group.

EXAMPLE 2.8.8. The quotient of the group of $n \times n$ matrices of determinant 1 over the real or complex numbers, by its center, is simple.

It is reported as of 1982 that all finite simple groups have been classified (the proof being some five thousand pages of research done by many group theorists). These include finite analogues of the groups in the last example, other matrix groups, subgroups of the symmetric group called ***alternating groups*** defined later, and certain other groups.

EXERCISES

Level 1

1. Prove the center of a group is closed under products and inverses (so is a semigroup).
2. Prove the center is a normal subgroup.
3. Find three subgroups of order 2 of the symmetric group of degree 3 from the multiplication table given in Example 2.5.3 (these will be $\{\{e, g\} : g^2 = e\}$).
4. Show none of the subgroups in the last exercise is normal.
5. Find all cosets of the subgroup $\{e, x\}$ in the symmetric group of degree 3. (How many will there be, by Theorem 2.5.1?)
6. Find two elements of the symmetric group of order 3 which generate the entire group. (Many pairs will do.)
7. What are some Sylow 2 and 3 subgroups of the symmetric group of degree 3?

Level 2

1. Prove that a group G of prime order p has no subgroup H except G, $\{e\}$ from the fact that $|H|$ divides $|G|$.
2. Prove that if $x \neq e$ in a group G of order p then the cyclic subgroup generated by x is all of G.
3. Using the last two exercises and a theorem in the last section prove a group of prime order p is isomorphic to $\mathbf{Z_p}$.
4. Prove that in a direct product $G \times H$ the sets $\{(e, g) : g \in G\}$ and $\{(e, h) : h \in H\}$ are normal subgroups.
5. Prove that if H is a subgroup of G and $|G|/|H| = 2$ then H is normal. For $x \notin H$ show $xH = G - H$ and $Hx = G - H$. For $x \in H$, $xH = Hx = H$.
6. Prove that if H, K are subgroups so is $H \cap K$.
7. What are the quotient groups for the normal subgroups in Exercise 4?

Level 3

1. What are all subgroups of $\mathbf{Z_p} \times \mathbf{Z_p}$ (note that except for G, $\{e\}$ they have order p and by Exercise 3 above are cyclic)?
2. Prove for subgroups H, K the set HK is a subgroup if and only if $HK = KH$.
3. Prove that if H is a subgroup and N is a normal subgroup $HN = NH$.
4. The quaternion group of order 8 has elements ± 1, $\pm i$, $\pm j$, $\pm k$ with multiplication $ij = k$, $ji = -k$, $kj = -i$, $ki = j$, $ik = -j$, signs treated as usual in algebra, and 1 the identity. Write out the multiplication table and find the center and all subgroups. They must have orders 1, 2, 4, 8. Orders 1, 8 are $\{e\}$, G. The order 2 subgroups are cyclic since 2 is prime. The order 4 subgroups are normal by Exercise 5 above.
5. Do the same for the dihedral group of order 8 which is the group of rotations and reflections of a square. It can also be represented by generators x (90° rotation) and y (reflection) with $x^4 = e$, $y^2 = e$, $yx = x^3y$. The

elements are $e, x, x^2, x^3, y, xy, x^2y, x^3y$ and products follow directly from these relations. Show that there exists a noncyclic subgroup of order 4 so that the dihedral and quaternion groups are not isomorphic. What is the quotient of this group by its center?

6. For any prime p there exists a nonabelian group of order p^3 with generators x, y, z and defining relations $x^p = 1$, $y^p = 1$, $z^p = 1$, $xy = yxz$, $zx = xz$, $zy = yz$. All elements have the form $x^n y^m z^r$ and products are given by $x^n y^m z^r x^a y^b z^c = x^{a+n} y^{b+m} z^{r+b+am}$. Find the center. Show every subgroup containing the center is normal. Find some normal subgroups of order p^2. *What is the quotient of this group by its center?*

2.9 HOMOMORPHISMS

A homomorphism of groups is a function preserving products.

DEFINITION 2.9.1. A function $f: G \to H$ is a *homomorphism* of groups if and only if $f(xy) = f(x)f(y)$ for all x, y in G.

As was earlier remarked homomorphisms that are not 1-to-1 can simplify the structure of a group. One-to-one homomorphisms can represent a group in terms of some other structure: a group can be regarded as a group of matrices or permutations.

EXAMPLE 2.9.1. If H is a subgroup of G the mapping $f(x) = x$ from H to G is a homomorphism.

EXAMPLE 2.9.2. The logarithm gives a group homomorphism from the positive real numbers under addition to the additive real numbers. This is expressed by $\log(xy) = \log x + \log y$.

EXAMPLE 2.9.3. For any constant c the mapping $f(x) = cx$ is a homomorphism from the additive real numbers to itself since $c(x + y) = cx + cy$.

EXAMPLE 2.9.4. The determinant is a homomorphism from $n \times n$ nonsingular matrices to nonzero real numbers, since $\det(XY) = \det(X)\det(Y)$.

EXAMPLE 2.9.5. For any commutative group and integer n the mapping $x \to x^n$ is a homomorphism of the group into itself since $(xy)^n = x^n y^n$ holds.

One-to-one onto homomorphisms are called ***isomorphisms***. An isomorphism between two groups means that it suffices to identify the structure of only one, since the other will have exactly the same structure. The logarithm is such an isomorphism and shows the group-theoretic properties of the positive real numbers under multiplication are just like the properties of the real numbers under addition.

One-to-one homomorphisms are called ***monomorphisms***, ***injections***, or ***embeddings***. Onto homomorphisms are called ***epimorphisms***, ***surjections***, or ***quotient maps***.

PROPOSITION 2.9.1. *If f is a homomorphism of groups, $f(e) = e$ and $f(x^{-1}) = (f(x))^{-1}$.*

Proof. We have $f(e)f(e) = f(ee) = f(e) = ef(e)$. Multiply by $f^{-1}(e)$ on the right, we have $f(e) = e$. We have $f(x)f(x^{-1}) = f(xx^{-1}) = f(e) = e = f(x)f^{-1}(x)$. Apply $f^{-1}(x)$ on the left. This completes the proof. □

Two of the most important aspects of homomorphisms are the kernel and image.

DEFINITION 2.9.2. The *kernel* of a homomorphism $f: G \to H$ is $\{x \in G: f(x) = e\}$.

DEFINITION 2.9.3. The *image* of a homomorphism $f: G \to H$ is $\{f(x): x \in G\}$.

EXAMPLE 2.9.6. If f is an isomorphism, its image is H and its kernel is $\{e\}$, since it is 1-to-1. If f sends all of G to e, its kernel is G and its image is e.

The image describes the result of applying the homomorphism. The kernel tells how far the map is from being 1-to-1, since $f(x) = f(y)$ if and only if $f(x)f^{-1}(y) = f(x)f(y^{-1}) = f(xy^{-1}) = e$, if and only if $xy^{-1} \in$ kernel.

EXAMPLE 2.9.7. For any normal subgroup N of a group G there is the mapping $x \to xN$ from $G \to G/N$. This is an epimorphism with kernel N.

PROPOSITION 2.9.2. *The image is always a subgroup and the kernel is always a normal subgroup.*

Proof. $f(xy) = f(x)f(y)$. If $f(x) = f(y) = e$ and $z \in G$ then $f(xy) = f(x)f(y) = e$ and $f(zxz^{-1}) = f(z)f(x)f(z^{-1}) = f(z)e(f(z))^{-1} = e$. □

The relationships between kernel and image can generally be expressed in a similar way to the epimorphism onto a quotient group.

THEOREM 2.9.3. *Let $f: G \to H$ be a homomorphism with kernel N. Then there is an isomomorphism g from G/N to the image of f, with the isomorphism described by $g(xN) = f(x)$.*

Proof. We show g does not depend on the choice of x, is a homomorphism, is onto, and is 1-to-1.

If $xN = yN$ then $x = yn$ for some $n \in N$. Then $f(x) = f(yn) = f(y)f(n) = f(y)e = f(y)$ since n is the kernel. This shows $g(xN)$ is the same for any choice of x.

We have $g(xNyN) = g(xyN) = f(xy) = f(x)f(y) = g(xN)g(yN)$. Therefore g is a homomorphism.

For $f(x)$ in the image, $f(x) = g(xN)$.

If $g(xN) = g(yN)$ then $f(x) = f(y)$, $f(xy^{-1}) = f(x)f^{-1}(y) = e$ so $xy^{-1} \in N$, $x \in Ny = yN$. Thus $xN = yN$. □

EXAMPLE 2.9.8. The function $|x|$ is a homomorphism from the nonzero real numbers under multiplication to the positive real numbers under multiplication. The kernel is ± 1. This gives an isomorphism from $\mathbf{R}\backslash\{0\}/\{x: |x| = 1\}$ onto $\mathbf{R}^+$ where $\mathbf{R}^+$ denotes the positive real numbers.

EXERCISES

Level 1

1. What is the image of the determinant map on nonsingular matrices? Give examples of elements of its kernel.
2. Consider the homomorphism $G \times H \to H$ such that $f(g, h)$ is h. What are its kernel and image?
3. The map in the above exercise gives an isomorphism between $(G \times H)/(G \times e)$ and what group, by the theorem? How does this partly justify the reason for the name 'quotient group'? (The order of the quotient group is another reason, as well as the division into cosets.)
5. Find a monomorphism from $\mathbf{Z_m}$ into the multiplicative group of complex numbers $\mathbf{C}$ (consider mth roots of unity).

Level 2

1. Prove any group homomorphism is the composition of an epimorphism and a monomorphism, using the theorem.
2. Show any group homomorphism $f: G \to H$ is the composition of a monomorphism and an epimorphism. Take as monomorphism $f_1: G \to G \times H$ where $f_1(g) = (g, f(g))$.
3. For any complex number x, define a multiplicative homomorphism f from the real numbers under addition into the complex numbers under addition, such that $f(1) = x$.
4. Let f from nonzero rationals under multiplication to the additive integers be defined as the power of 2 occurring in the rational number. Give the image and kernel of f.
5. Prove the intersection of two normal subgroups is normal.
6. Let $f_1: G \to H_1$ and $f_2: G \to H_2$ be homomorphisms with kernels N_1, N_2. Let $f(x) = (f_1(x), f_2(x)) \in H_1 \times H_2$. Show f is a homomorphism. What is its kernel?

Level 3

1. Prove that if $N \subset K \subset G$ and N, K are normal subgroups, $\dfrac{G/N}{K/N}$ is isomorphic to G/K. (Define a composite epimorphism $G \to G/N \to \dfrac{G/N}{G/K}$ and show the kernel is precisely K.) This is a basic theorem of group theory.
2. Show that the transformations $f(x) = ax + b$, $a \neq 0$ have a normal subgroup of transformations of the form $f(x) = x + b$. Show the quotient group is isomorphic to the multiplicative nonzero real numbers.
3. There is a homomorphism from the additive real numbers to the multiplicative complex numbers, described by $f(x) = \cos x + i \sin x$ (that is, e^{ix}). Find its kernel and image.
4. Prove that if $H \subset G$ is a subgroup and $N \subset G$ is a normal subgroup, $H \cap N$ is normal in H. Prove HN/N is isomorphic to $H/H \cap N$. (Map H into HN/N by $f(x) = xN$. Show the image is all of HN/N and the kernel is $H \cap N$.) This is a basic theorem of group theory.

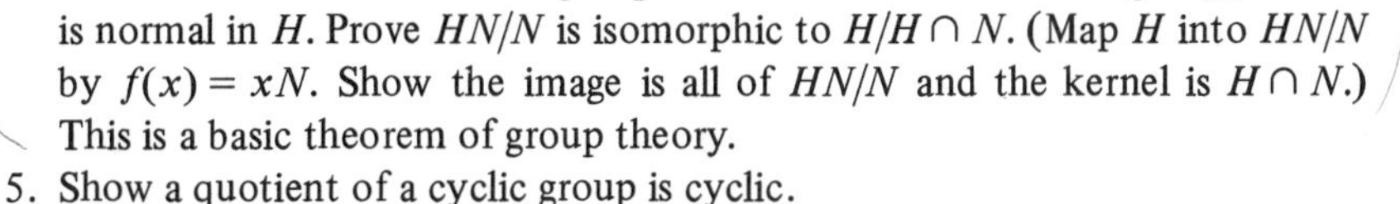

5. Show a quotient of a cyclic group is cyclic.
6. Show the set $N_m = \{mx : x \text{ an integer}\}$ is a normal subgroup of $\mathbf{Z}$. The quotient group must be cyclic since $\mathbf{Z}$ is. What is the quotient group isomorphic to?

2.10 PERMUTATION GROUPS

A ***permutation*** f on $\{1, 2, \ldots, n\}$ is a 1-to-1, onto transformation from the set $\{1, 2, \ldots, n\}$ to itself. Permutations (and transformations) are frequently written in the form

$$\begin{pmatrix} 1 & 2 & \ldots & n \\ f(1) & f(2) & \ldots & f(n) \end{pmatrix}$$

That is, the top row is always $1, 2, \ldots n$. Underneath each number x is the value $f(x)$.

EXAMPLE 2.10.1. Let f be the permutation $f(1) = 2$, $f(2) = 3$, $f(3) = 1$. Then f is

$$\begin{pmatrix} 1 & 2 & 3 \\ 2 & 3 & 1 \end{pmatrix}$$

Permutations here will be multiplied as if composing them on the right. That is, $(x)fg = ((x)f)g$. Apply f to x, then apply g to the result $f(x)$. This can be done quickly using the double-row notation.

EXAMPLE 2.10.2. The product

$$\begin{pmatrix} 1 & 2 & 3 & 4 \\ 2 & 3 & 4 & 1 \end{pmatrix} \begin{pmatrix} 1 & 2 & 3 & 4 \\ 4 & 3 & 2 & 1 \end{pmatrix}$$

is $1 \to 2 \to 3, 2 \to 3 \to 2, 3 \to 4 \to 1, 4 \to 1 \to 4$. This is

$$\begin{pmatrix} 1 & 2 & 3 & 4 \\ 3 & 2 & 1 & 4 \end{pmatrix}$$

To find such a product, for each x look underneath x in the left-hand permutation. Find that same number on top of the right-hand permutation. What is under it goes below x in the result.

A second common way of writing permutations is as products of cycles. A ***cycle*** is a sequence like $1 \to 4, 4 \to 2, 2 \to 3, 3 \to 1$ which runs through a set of numbers once, and ends where it began, the function being applied each time to the result at the last stage. A cycle is a set of numbers x, $f(x)$, $f^2(x), \ldots, f^k(x) = x$. Numbers not in the cycle are left unchanged by it. The above cycle would actually be written as (1423).

DEFINITION 2.10.1. For distinct x_i, $(x_1 x_2 \ldots x_k)$ is the permutation f such that $f(x_1) = x_2$, $f(x_2) = x_3$, in general $f(x_i) = x_{i+1}$, and $f(x_k) = x_1$. For $x \notin \{x_1, x_2, \ldots, x_k\}$, $f(x) = x$. Such a permutation is called a *cycle* or *k-cycle*.

EXAMPLE 2.10.3. The cycle (123) is

$$\begin{pmatrix} 1 & 2 & 3 \\ 2 & 3 & 1 \end{pmatrix}$$

This notation is not unique: $(123) = (231) = (312)$.

Two cycles are ***disjoint*** if and only if the sets $\{x_1, x_2, \ldots, x_k\}$ are disjoint. The notation (x) represents the identity permutation and is sometimes called a ***1-cycle***.

Cycle notation is more compact and tells more about the permutation than the double-row notation. But multiplication of cycles is less simple.

THEOREM 2.10.1. *Every permutation is a product of disjoint cycles. Disjoint cycles commute.*

Proof. Let p be a permutation on $\{1, 2, \ldots, n\}$. Consider the relation $\{(x, y)$: $x = (y)p^i$ for some integer $i\}$. This is an equivalence relation since $x = (x)p^0$ and if $x = (y)p^i$ then $(x)p^{-i} = y$ and if $x = (y)p^i$ and $y = (z)p^j$ then $zp^{j+i} = yp^i = x$. To each equivalence class we will associate a cycle of p. Namely for an integer x, let n be the least positive integer such that $(x)p^n = x$.

Then for $0 < s < t < n$; if $(x)p^s = (x)p^t$ then $(x) = (x)p^s p^{-s} = (x)p^t p^{-s} = (x)p^{t-s}$. This contradicts the assumption on n. So $x, (x)p, (x)p^2, \ldots, (x)p^{n-1}$ are distinct. We have a cycle $(x\ (x)p\ (x)p^2 \ldots (x)p^{n-1})$. The product of all such cycles is p. This proves the first assertion.

Let f, g be two disjoint cycles. Then for all x either $(x)f = x$ or $(x)g = x$. Moreover if $x \neq (x)f$ then $(x)f \neq (x)f^2$. Suppose $x \neq (x)f$. Then $(x)g = x$ and $((x)f)g = (x)f$. So $(x)gf = (x)f$ and $(x)fg = (x)f$. Likewise if $x \neq (x)g$, $(x)gf = (x)fg$. Suppose $x = (x)f$ and $x = (x)g$. Thus $(x)gf = x = (x)fg$. This proves the last statement. □

EXAMPLE 2.10.4. If

$$\begin{pmatrix} 1 & 2 & 3 & 4 & 5 & 6 & 7 & 8 & 9 \\ 4 & 5 & 7 & 9 & 2 & 3 & 6 & 1 & 8 \end{pmatrix}$$

we have $1 \to 4 \to 9 \to 8 \to 1$, $2 \to 5 \to 2$, $3 \to 7 \to 6 \to 3$. Thus the permutation equals (1498) (25) (376).

Permutations in cycle form can be multiplied by tracing the image of $1, 2, \ldots, n$. Afterwards the cycles of the product must usually be found.

EXAMPLE 2.10.5. Let $p = (123)(34)$. In this product $1 \to 2 \to 2$, $2 \to 3 \to 4$, $3 \to 1 \to 1$, $4 \to 4 \to 3$. So the product is

$$\begin{pmatrix} 1 & 2 & 3 & 4 \\ 2 & 4 & 1 & 3 \end{pmatrix}$$

This is (1243).

DEFINITION 2.10.2. The *order* of an element $x \in G$ is the least positive integer n such that $x^n = e$.

EXAMPLE 2.10.6. The order of the permutation

$$\begin{pmatrix} 1 & 2 & 3 \\ 2 & 3 & 1 \end{pmatrix}$$

is 3.

The order of an element is the same as the order of the cyclic subgroup it generates, since that subgroup is $\{e, x, x^2, \ldots, x^{n-1}\}$. Thus the order of any element of a group divides the order of the group, since the order of a subgroup divides the order of a group.

PROPOSITION 2.10.2. *The order of a k-cycle is k. The order of a product of disjoint cycles is the least common multiple of their orders.*

Proof. If f is a k-cycle, it can be seen that $xf^k = x$ and that if x is in the cycle, k is the lowest power with this property. So the order of f is k.

Let f be a product $z_1 z_2 \ldots z_r$ of disjoint cycles. Then since the cycles commute, $f^m = z_1^m z_2^m \ldots z_r^m$. This will be the identity if and only if $z_1^m = e$, $z_2^m = e, \ldots, z_r^m = e$, that is, if m is divisible by the order of each cycle. The last statement of the proposition follows from this. □

EXAMPLE 2.10.7. The order of (1467) (2398510) is the least common multiple of 4, 6, that is, 12.

PROPOSITION 2.10.3. *Let p be a permutation and $z = (x_1 x_2 \ldots x_k)$ a k-cycle. Then pzp^{-1} is the k-cycle $(y_1 y_2 \ldots y_k)$ such that $y_i = p^{-1}(x_i)$ for each i.*

Proof. Under pzp^{-1} the element $p^{-1}(x_i)$ is sent to x_i then to x_{i+1} then to $p^{-1}(x_{i+1})$. So $y_i \to y_{i+1}$. Also $y_k \to y_1$. And elements v not of the form $p^{-1}(x_i)$ have $v \to p(v) \to p(v) \to v$. □

For products of cycles pzp^{-1} can be computed in the same way since $pxp^{-1}pyp^{-1} = pxyp^{-1}$. Here p^{-1} was defined after Proposition 1.3.3.

EXAMPLE 2.10.8. $(123)(12)(123)^{-1} = (31)$.

EXERCISES

Level 1

1. Multiply

$$\begin{pmatrix} 1 & 2 & 3 & 4 \\ 2 & 3 & 4 & 1 \end{pmatrix} \begin{pmatrix} 1 & 2 & 3 & 4 \\ 3 & 1 & 4 & 2 \end{pmatrix}$$

2. Find all powers of

$$\begin{pmatrix} 1 & 2 & 3 & 4 & 5 \\ 2 & 3 & 1 & 5 & 4 \end{pmatrix}$$

3. Write

$$\begin{pmatrix} 1 & 2 & 3 & 4 & 5 \\ 2 & 3 & 1 & 5 & 4 \end{pmatrix}$$

in cycle form.

4. Write the multiplication of the set of permutations e, (12) (34), (13) (24), (14) (23). Show it forms a commutative subgroup. It is called the ***Klein 4-group***.
5. What is (12) (23)? (12) (23) (34)?

Level 2

1. Find a permutation of degree 7 but of order 10. (It will have a 5-cycle and a 2-cycle.) What is the largest order for degree 8?
2. What is the product $(xy)(yz)$? Here x, y, z are distinct.
3. What is the general product $(x_1 x_2)(x_2 x_3) \ldots (x_{k-1} x_k)$? Here all x_i are distinct. Tell why this implies that 2-cycles, called ***transpositions***, generate the symmetric group.
4. Can a power of a cycle have more than one cycle? When? Consider (1234) and (12345).
5. Tell how to obtain the inverse of a permutation written in cycle form.

Level 3

1. Find $(12 \ldots n)(23)(12 \ldots n)^{-1}$. Note $(12 \ldots n)^{-1} = (n \ldots 21)$. Note for $x \neq 1, 2, n$; $x \to x+1 \to x+1 \to x$, $1 \to 2 \to 3 \to 2$, $2 \to 3 \to 2 \to 1$, $n \to 1 \to 1 \to n$.
2. Find $(12 \ldots n)(i+1\ i+2)(12 \ldots n)^{-1}$ for $i = 1, 2, \ldots, n-2$.
3. Generalize the formula $(54)(43)(32)(12)(23)(34)(45) = (15)$. Thus show all transpositions are products of $(i\ i+1)$.
4. Combine the answers to 2, 3 of Level 3 and 3 of Level 2 to find two permutations x, y which generate the entire symmetric group.
5. Show that all transformations are generated by the two permutations in the above exercise together with the transformation $f(x) = x$, $x \neq 1$, $f(1) = 2$. First obtain all transformations with image size $n-1$ as pfq for permutations p, q. Then show any transformation with image size k, $k < n-1$ can be factored as a product of transformations with image size $k+1$.

2.11 SYSTEMS OF DISTINCT REPRESENTATIVES AND FLOWS ON NETWORKS

Systems of distinct representatives are solutions to a problem such as this: *Suppose a workshop has n rooms. Each room i can be used for a specific set of tasks S_i which must be done. How can we assign a task to each room, with the understanding that all tasks are distinct?*

EXAMPLE 2.11.1. Suppose tasks 1, 2 can be performed in room 1, tasks 2, 3 in room 2, and task 2 only in room 3. Then there is only one possible assignment: task 1 to room 1, task 3 to room 2, task 2 to room 3.

Many combinatorial problems reduce to finding a system of distinct representatives.

DEFINITION 2.11.1. Let $U_1, U_2, \ldots, U_n$ be subsets of a set U. A *system of distinct representatives* (SDR) is a sequence $u_1, u_2, \ldots, u_n$ such that the u_i are distinct and $u_i \in U_i$ for all i.

Systems of distinct representatives can conveniently be studied by means of a type of matrix defined below.

DEFINITION 2.11.2. A *permutation matrix* P is a Boolean matrix (or $(0,1)$-matrix) which is the matrix of a permutation π on $\{1, 2, \ldots, n\}$, i.e. a 1-to-1 onto function. Thus $p_{ij} = 1$ if and only if $(i)\pi = j$.

EXAMPLE 2.11.2. The matrix of the permutation

$$\begin{pmatrix} 1 & 2 & 3 & 4 \\ 2 & 3 & 1 & 4 \end{pmatrix}$$

is

$$\begin{bmatrix} 0 & 1 & 0 & 0 \\ 0 & 0 & 1 & 0 \\ 1 & 0 & 0 & 0 \\ 0 & 0 & 0 & 1 \end{bmatrix}$$

PROPOSITION 2.11.1. *A Boolean matrix (or $(0, 1)$-matrix) is a permutation matrix if and only if it has exactly one 1 in each row and column.*

Proof. A permutation matrix has a 1 on place $(i)\pi$ in row i and place $(j)\pi^{-1}$ in column j. Conversely if A has exactly one 1 in each row and column, define π by $a_{i(i)\pi} = 1$. Then π must be a 1-to-1 else some column would have two ones. So it is a permutation. □

DEFINITION 2.11.3. A Boolean matrix (or $(0,1)$-matrix) A is a *Hall matrix* if and only if there exists a permutation matrix P such that $P \leqslant A$.

EXAMPLE 2.11.3. There is no permutation matrix $P \leqslant A$ for

$$A = \begin{bmatrix} 0 & 1 & 1 \\ 1 & 0 & 0 \\ 1 & 0 & 0 \end{bmatrix}$$

Therefore it is not a Hall matrix.

Hall matrices will here be considered usually as Boolean matrices, since they form a semigroup under Boolean matrix multiplication. Treatment as a Boolean matrix makes no difference except that the operations are Boolean, but for a (0, 1)-matrix they are the usual arithmetic operations in the real number field.

For a system of distinct representatives we form a matrix A. Let $|U| = m$ and label the elements of U as $u_1, u_2, \ldots, u_m$. Form an $n \times m$ matrix A by $a_{ij} = 0$ if $u_j \notin U_i$ and $a_{ij} = 1$ if $u_j \in U_i$. Thus the ith row of A has ones for precisely those places corresponding to the members of U_i.

EXAMPLE 2.11.4. Let $U_1 = \{1, 2, 4\}$, $U_2 = \{3, 4\}$, $U_3 = \{1, 3\}$, $U_4 = \{1, 4\}$. Then

$$\begin{bmatrix} 1 & 1 & 0 & 1 \\ 0 & 0 & 1 & 1 \\ 1 & 0 & 1 & 0 \\ 1 & 0 & 0 & 1 \end{bmatrix}$$

There are two SDRs. One is $u_1 = 2, u_2 = 3, u_3 = 1, u_4 = 4$.

In a square Boolean matrix A, an SDR corresponds to a set of 1 entries of A such that there is precisely one 1 entry in each row and column, that is, a permutation matrix $P \leqslant A$.

EXAMPLE 2.11.5. The SDR in Example 2.11.4 corresponds to the permutation matrix of Example 2.11.2.

If $m < n$ the set U can have no SDR. If $m > n$ we take a square $m \times m$ matrix by simply adding sets $U_{n+1}, U_{n+2}, \ldots, U_m$ all equal to U. An SDR exists in the new system if and only if one exists in the old system.

By the above remarks, a square Boolean matrix is a Hall matrix if and only if the corresponding system of sets has an SDR. There exists a famous, but somewhat difficult to prove theorem characterizing when a matrix is a Hall matrix.

We will prove this using an ***algorithm of Ford*** and ***Fulkerson*** in network theory. The algorithm will not be proved in detail here but a proof is sketched.

DEFINITION 2.11.4. An $r \times s$ *rectangle of zeros* (*ones*) in a Boolean matrix (or (0, 1)-matrix) A is a set $R \times S \subset \{1, 2, \ldots, n\} \times \{1, 2, \ldots, n\}$ such that $a_{ij} = 0$ ($a_{ij} = 1$) for $i \in R$, $j \in S$ and $|R| = r$ and $|S| = s$.

EXAMPLE 2.11.6. In the Boolean matrix

$$\begin{bmatrix} 0 & 1 & 0 \\ 1 & 1 & 1 \\ 0 & 1 & 0 \end{bmatrix}$$

$\{1, 3\} \times \{1, 3\}$ gives a 2×2 rectangle of zeros.

DEFINITION 2.11.5. A matrix A is *full* if it has no $r \times s$ rectangle of zeros where $r, s \neq 0$ and $r + s = n + 1$.

EXAMPLE 2.11.7. If A is

$$\begin{bmatrix} 1 & 0 & 1 & 1 \\ 0 & 1 & 1 & 1 \\ 1 & 0 & 1 & 1 \\ 1 & 1 & 1 & 0 \end{bmatrix}$$

then A is a full matrix.

A ***network*** is a directed graph having a vertex x_0 (***source***) such that every edge from x_0 is directed outwards and a vertex z (***sink***) such that every edge from z goes to z. Each edge e is assigned an integer $c(e)$ called its *capacity*. A flow consists of an assignment of integers $f(e)$ to edges such that (1) at each vertex except x_0, z the sum of all incoming values $f(e)$ equals the sum of all outgoing values $f(e)$, (2) $f(e) \leqslant c(e)$. The value of the flow is the sum $f(e)$ over all incoming edges at z. This equals the sum over outgoing edges at x_0. One can think of a flow as oil flowing through a collection of pipes from x_0 to z or goods being shipped along a network of roads from x_0 to z.

EXAMPLE 2.11.8. For this network

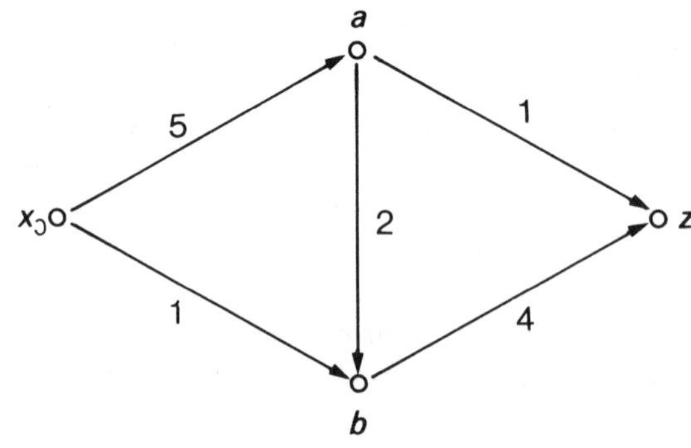

there exists the following flow

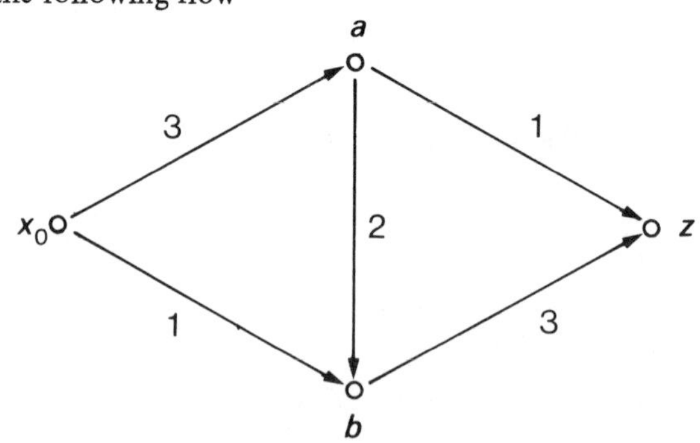

The Ford-Fulkerson algorithm starts with a flow of zeros. It then increases the flow if possible by the following method, by one unit. We form sets S_k where $S_0 = \{x_0\}$. We put a point x in S_k if and only if it is not in

$$\bigcup_{i=0}^{k-1} S_i$$

and either (1) there exists an edge from a point $u \in S_{k-1}$ to x which is not flowing at full capacity $c(e)$ or (2) there exists an edge from x to a point $u \in S_{k-1}$ having positive flow. Continue as long as possible. If z belongs to some S_k then we can increase the flow by one unit. Trace a path $x_0, x_1, x_2, \ldots, z$, where $x_i \in S_i$. This can be done going backwards. For each edge of type (1) in the path increase the flow by one. For each edge of type (2) reduce the flow by one. This in effect adds 1 unit of flow along the path from x_0 to z so it increases the flow by 1.

EXAMPLE 2.11.9. For the preceding network start with a flow of zero. Now form sets $S_1 = \{a, b\}$, $S_2 = \{z\}$. Thus we can increase the flow by 1. One path from x_0 to z is $x_0 \to b \to z$. So increase the flow along this path.

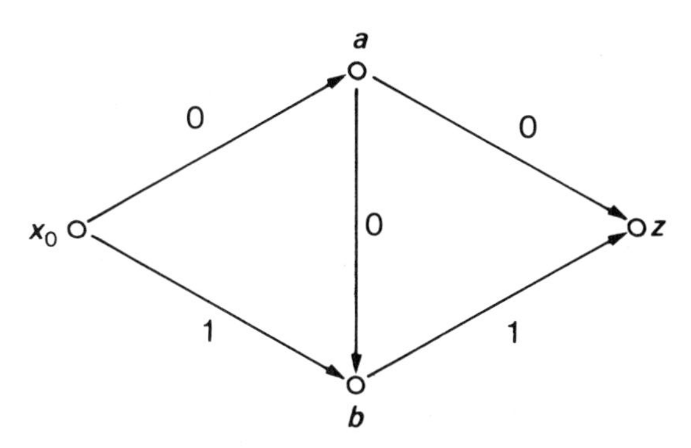

Now repeat. This time $S_1 = \{a\}$, $S_2 = \{b, z\}$. So increase the flow by 1 along the path x_0 to a to z.

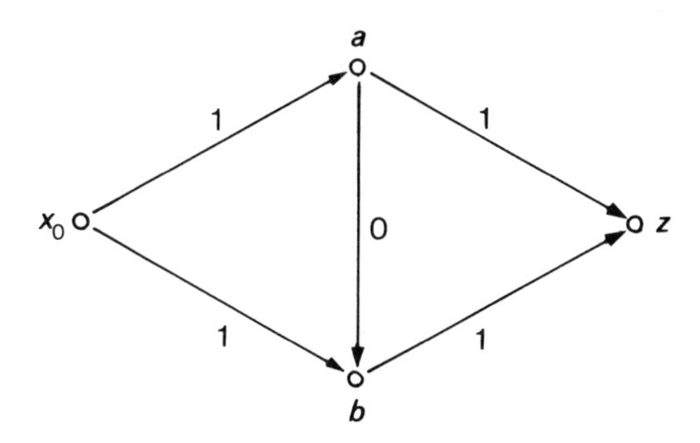

Repeat once more $S_1 = \{a\}$, $S_2 = \{b\}$, $S_3 = \{z\}$. Increase the flow by 1 along this path

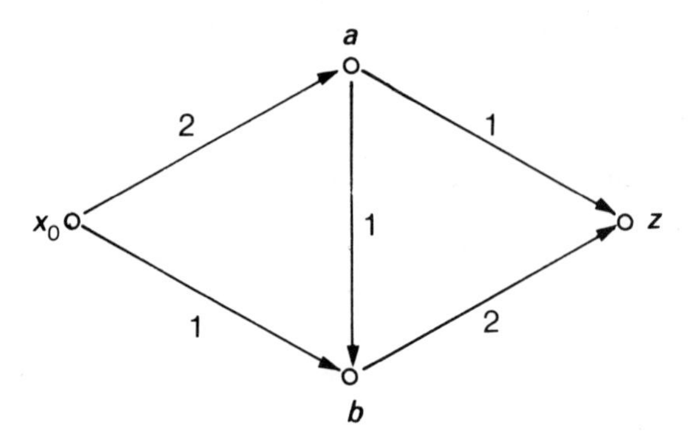

Finally again increase the flow by 1 along the same path. This gives the flow of the previous example.

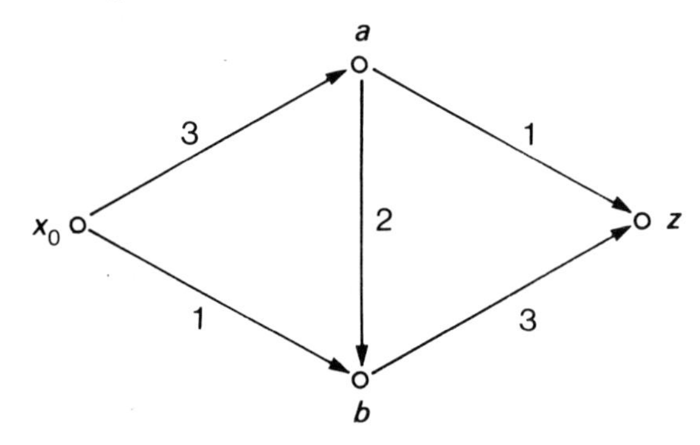

This is maximal since all edges from $\{x_0, a\}$ to $\{b, z\}$ are at full capacity.

EXAMPLE 2.11.10. Consider this network

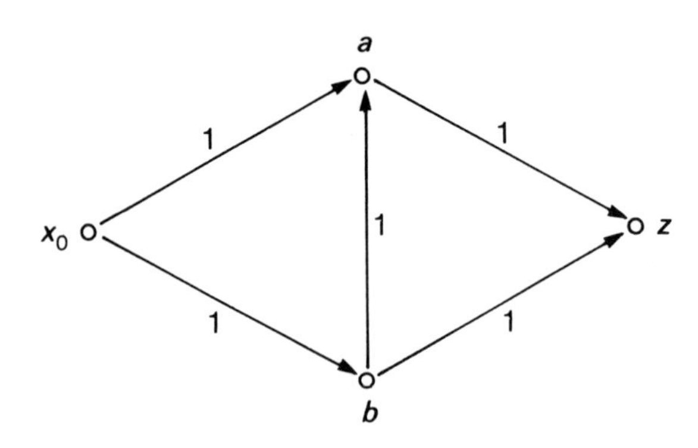

Suppose the flow

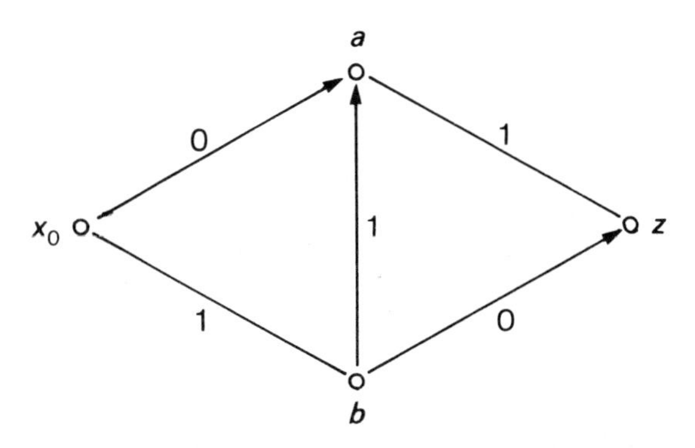

existed and it is desired to increase this flow by 1. Then form $S_1 = \{a\}$, $S_2 = \{b\}$, $S_3 = \{z\}$. This is an example where an edge of type (2) is used. Increase the flow by 1 unit from x_0 to a, reduce it by 1 from a to b, increase it by 1 from b to z.

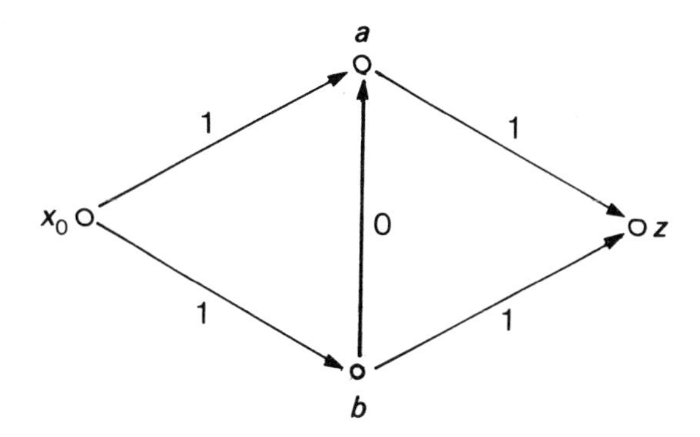

This is maximal since all edges out of x_0 are at full capacity.

THEOREM 2.11.2. (Ford and Fulkerson.) *If v is the maximum value of a flow from x_0 to z then there exists a set U of vertices such that $x_0 \in U$, $z \notin U$ and the total capacity of all edges from U to $\tilde{U}$ is v.*

Proof. The value of the flow cannot exceed the total capacity of edges from U to $\tilde{U}$ since, intuitively speaking all material flowing from x to z must go through one of these edges. In fact it can be shown that the value of the flow equals the amount flowing from U to $\tilde{U}$.

Suppose the flow is maximal. Then in the algorithm the sets S_k do not include z. Let $U = \cup S_i$. Then $x_0 \in U$, $z \notin U$. Moreover there exists no edge from U to $\tilde{U}$ not at full capacity and no positive flow from U to $\tilde{U}$ or there would be an additional set S_k. So the total capacity from U to $\tilde{U}$ equals the amount actually flowing which is v. □

THEOREM 2.11.3. (Hall and Koenig.) *A (0, 1)-matrix (or Boolean matrix) is not a Hall matrix if and only if it is full.*

Proof. No permutation matrix can have an $r \times s$ rectangle of zeros, where $r + s > n$, since $p(\tilde{R})$ must be onto S, but $|\tilde{R}| = n - r < s = |S|$, if p is the permutation and $R \times S$ is the rectangle of zeros.

Suppose A is not a Hall matrix. Draw a graph with vertices $x_0, y_1, y_2, \ldots, y_n, w_1, w_2, \ldots, w_n, z$ having directed edges from x_0 to y_i for each i, w_i to z for each i, and an edge from y_i to w_j if and only if $a_{ij} = 1$. Then if P is a permutation matrix, $P \leqslant A$ and p is the permutation whose matrix is P, a flow on the graph is defined by $g(x_0, y_i) = 1$, $g(y_i, w_{p(i)}) = 1$, $g(w_i, z) = 1$ and g is zero on all other edges. Here all edges have been assigned capacity 1. Conversely a flow of value n arises from a unique SDR. It follows from the Ford-Fulkerson Algorithm that if no such flow exists there is some set of vertices T such that at most $n - 1$ edges lead from T to $\tilde{T}$, and $x_0 \in T$, $z \in \tilde{T}$.

Let $R = \{y : y \in T\}$ and $S = \{w_j$ is not connected to a $y_i \in T\}$, $|R| = r$, $|S| = s$. There are at least $n - r$ edges from x_0 to members of y_i not in T. For each element w_j not in S there is at least one edge from an element y_i of R to it. If $w_j \notin T$ this counts as one of the edges from T to $\tilde{T}$. Else we have an edge from w_j to z which is from T to $\tilde{T}$. This gives at least $n - r + n - s$ edges T to $\tilde{T}$. So $n - r + n - s \leqslant n - 1$. So $r + s \geqslant n + 1$. So if no flow exists, R, S give a rectangle of zeros with $r + s > n$. □

EXAMPLE 2.11.11. This matrix has a 3×2 rectangle of zeros. So it is not a Hall matrix.

$$\begin{bmatrix} 1 & 1 & 1 & 1 \\ 1 & 1 & 0 & 0 \\ 1 & 1 & 0 & 0 \\ 1 & 1 & 0 & 0 \end{bmatrix}$$

EXERCISES

Level 1

1. Find an SDR for $U_1 = \{1, 3\}$, $U_2 = \{2, 3\}$, $U_3 = \{2, 4\}$, $U_4 = \{1, 4\}$.
2. Find a permutation matrix $P \leqslant A$.

$$\begin{bmatrix} 0 & 1 & 0 & 1 \\ 0 & 0 & 1 & 0 \\ 0 & 1 & 1 & 1 \\ 1 & 0 & 1 & 0 \end{bmatrix}$$

3. Find a permutation matrix $P \leqslant A$.

$$\begin{bmatrix} 0 & 1 & 1 & 1 \\ 1 & 0 & 0 & 1 \\ 1 & 0 & 1 & 0 \\ 1 & 1 & 0 & 0 \end{bmatrix}$$

4. Find a rectangle of zeros with $r + s > n$.

$$\begin{bmatrix} 1 & 1 & 1 & 1 & 1 \\ 0 & 1 & 0 & 0 & 0 \\ 0 & 0 & 1 & 0 & 0 \\ 1 & 1 & 1 & 1 & 0 \\ 0 & 0 & 1 & 0 & 0 \\ 1 & 1 & 1 & 0 & 1 \end{bmatrix}$$

5. Find a maximal flow in this network.

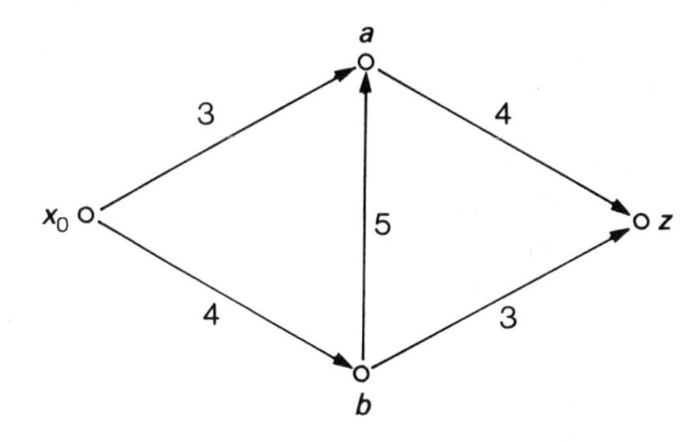

Level 2

1. Find all SDRs for this system $\{\{1,2\}, \{2,3\}, \{3,4\}, \{4,5\}, \{1,5\}\}$. Generalize to any n.
2. Suppose a (0, 1)-matrix has at least k rows with at most k ones for each $k \leqslant n$. Show the system has at most one SDR.
3. Find all permutation matrices P less than or equal to this matrix.

$$\begin{bmatrix} 0 & 1 & 1 & 1 & 1 \\ 1 & 0 & 0 & 0 & 1 \\ 1 & 0 & 0 & 1 & 0 \\ 1 & 0 & 1 & 0 & 0 \\ 1 & 1 & 0 & 0 & 0 \end{bmatrix}$$

4. The ***permanent*** of a matrix M is the sum over all permutations π of $m_{1(1)\pi} m_{2(2)\pi} \cdots m_{n(n)\pi}$. Thus it is the same expression as the determinant except that there are no negative signs. Show that for a (0, 1)-matrix A the permanent is the number of SDRs in the corresponding system.
5. A matrix is said to be *doubly stochastic* if it is nonnegative and all rows and columns have sum 1. Give examples of doubly stochastic $n \times n$ matrices with permanents precisely 1 and precisely

$$\frac{n!}{n^n}.$$

These are known to be the upper and lower bounds (the latter, the *van der Waerden conjecture,* was proved only recently).

Level 3

1. Show a square matrix having exactly k ones in each row and column must be a Hall matrix (show that an $R \times S$ rectangle of zeros with $|R| + |S| = n + 1$ would make this row and column sum condition false).
2. A Boolean matrix is k-*decomposable* if and only if it has a rectangle of zeros with $r + s = n - k + 1$. Else it is called k-*indecomposable*. Show a Boolean matrix A is k-indecomposable if and only if for every Boolean vector $v \neq 0$ either vA has no zeros or has at least k zeros fewer than v.
3. Using Exercise 2 show the Boolean product of a k-indecomposable and a p-indecomposable matrix is at least $(p + k)$-indecomposable. Thus for $k \geqslant 0$ the set of k-indecomposable matrices forms a semigroup.
4. Prove a Boolean matrix A is a Hall matrix if and only if MA has no more zeros than M, for any Boolean matrix M. Exercise 2 may be used row by row.
5. Vertices x, y of a directed graph are said to lie in the same k-***connected component*** if there exist flows of value k from x to y and from y to x. Here any ingoing edges to the source and any outgoing edges to the sink are disregarded. All edges have capacity one. Using the theorems of this section prove this is an equivalence relation.

2.12 ORBITS AND CONJUGATIONS

An automorphism, as mentioned earlier, is an isomorphism from a group to itself. The identity mapping is always one automorphism.

Every group of order greater than two has at least one other automorphism. If the group is abelian and not all elements are of order two, then $x \to -x$ is an automorphism which is not the identity. If all elements have order two and the group is commutative, then any permutation of a minimum generating set extends to an automorphism if its order exceeds two. Nonabelian groups have what is called ***inner automorphisms*** (and frequently others).

DEFINITION 2.12.1. Let G be a group and $g \in G$. Let C_g be the function from G to G such that $C_g(x) = gxg^{-1}$. Then C_g is called a *conjugation,* or *inner automorphism*.

EXAMPLE 2.12.1. In an abelian group $C_g(x) = gxg^{-1} = gg^{-1}x = x$.

EXAMPLE 2.12.2. A conjugation by a nonsingular matrix replaces a matrix by a matrix similar to that matrix.

DEFINITION 2.12.2. Two elements x, y of a group G are *conjugate* if and only if there exists $g \in G$ such that $y = gxg^{-1}$.

EXAMPLE 2.12.3 In Example 2.8.3, x, y, z are all conjugate, since $y = bxb^{-1}$ and $z = axa^{-1}$.

THEOREM 2.12.1.

(*1*) $C_g(xy) = C_g(x)C_g(y)$.
(*2*) $C_g C_h(x) = C_{gh}(x)$.
(*3*) C_g *is an isomorphism* $G \to G$.
(*4*) *Conjugacy is an equivalence relation.*

Proof. $C_g(xy) = gxyg^{-1}$. $C_g(x)\,C_g(y) = gxg^{-1}gyg^{-1} = gxeyg^{-1} = gxyg^{-1}$. This proves (1).

$C_g(C_h(x)) = C_g(hxh^{-1}) = ghxh^{-1}g^{-1} = (gh)x(gh)^{-1}$. This proves (2).

Let g, h be inverses of each other. Then C_g and C_h are inverse functions of each other. So each is 1-to-1 and onto. This proves (3).

For reflexivity, $x = exe^{-1}$. If $x = gyg^{-1}$ then $y = (g^{-1})x(g^{-1})^{-1}$. If $x = C_g(y)$ and $y = C_h(z)$, then $x = C_{gh}(z)$. This proves conjugacy is an equivalence relation. □

THEOREM 2.12.2. *Two permutations are in the same conjugacy class if and only if for* $k = 1$ *to* n *they have the same number of* k*-cycles when expressed as products of disjoint cycles.*

Proof. By Proposition 2.10.3, two conjugate permutations do have the same number of k-cycles for each k. Conversely let f and g be products of disjoint cycles such that the same number of k-cycles occur in both, for each k. Let p send the elements of each cycle of f, in order, to the elements of a cycle of g having the same length. Let p be an arbitrary isomorphism on any remaining elements of $\{1, 2, \ldots, n\}$. Then by Proposition 2.10.3, $pgp^{-1} = f$. □

This means that the number of conjugacy classes of the symmetric group of degree n equals the number of partitions of the integer n, that is the number of distinct sets of positive integers whose sum is n.

The degree of a permutation on $\{1, 2, \ldots, n\}$ is n. This should be distinguished from the order.

EXAMPLE 2.12.4. The symmetric group of degree 3 has three conjugacy classes. Representatives are $e = (1)(2)(3), (12), (3), (123)$. The three partitions of 3 corresponding to these are

$$3 = 1 + 1 + 1$$

$$3 + 2 + 1$$

$$3 = 3$$

A group of permutations on a set is ***transitive*** if it takes every member of the set onto every other. For instance, the group of rigid motions in plane geometry (generated by translations, rotations, and reflections) is transitive on points since any point can be translated to any other, on lines since any line can be translated and rotated to any other, on right angles since any right angle can be translated and rotated to coincide with any other. It is not transitive on all triangles since noncongruent triangles cannot be made to coincide by rigid motions.

A group is ***imprimitive*** if it moves the elements in sets such that each set is always sent onto another set. For example the additive group of integers acting on itself is imprimitive, since the odd integers to precisely the odd integers or precisely the even integers, on addition by any number. Let $\mathbf{Z^0}$, $\mathbf{Z^E}$ denote the sets of odd integers, and of even integers, respectively. For an integer x and set S let $x + S$ denote $\{x + y : y \in S\}$. Then for any x either $x + \mathbf{Z^0} = \mathbf{Z^0}$ or $x + \mathbf{Z^0} = \mathbf{Z^E}$.

Another way to say this is that it is imprimitive if it preserves an equivalence relation other than the identity or universal relation. These two are always preserved. It will then send any equivalence class to some equivalence class.

DEFINITION 2.12.3. Let G be a finite group of permutations of a set X. Then G is said to be ***transitive*** if for all x, y in X there exists $g \in G$ such that $(x)g = y$, and G is said to be *primitive* if there does not exist an equivalence relation R such that $(x, y) \in R$ if and only if $((x)g, (y)g) \in R$, other than the identity and the universal relation.

EXAMPLE 2.12.5. The cyclic group generated by $(12)(34)$ is not transitive, since $(1)g = 3$ never holds.

EXAMPLE 2.12.6. The cyclic group generated by (1234) is transitive, but not primitive, since it preserves the partition $\{\{1, 3\}, \{2, 4\}\}$. That is the set $\{1, 3\}$ is either mapped to itself or to $\{2, 4\}$, never to some set that overlaps both.

EXAMPLE 2.12.7. The group of permutations e, (123), (132) on $\{1, 2, 3\}$ is transitive and primitive.

If a group is not transitive there will be more than one subset, called an ***orbit***, restricted to which it is transitive.

DEFINITION 2.12.4. Let G be a group of permutations on the set X. Then x and y are said to belong to the same *orbit* if and only if $(x)g = y$ for some g. The isotropy subgroup of x is $\{g : (x)g = x\}$.

The idea is that under the group action any element stays within its own orbit, eventually reaching all points of the orbit, but does not go outside that orbit.

EXAMPLE 2.12.8. The group e, (12) (34) has two orbits $\{1, 2\}$ and $\{3, 4\}$. The isotropy subgroup of 1 is $\{e\}$. Under the action $1 \to 2 \to 1$ and $3 \to 4 \to 3$. The number 1 will never be sent to 3.

The orbits of a cyclic group are the sets for the cycles of its generator.

THEOREM 2.12.3. *The relation of belonging to the same orbit is an equivalence relation. A group is transitive if and only if there is exactly one orbit. If a group is primitive, it must be transitive. If a group is transitive but not primitive, all equivalence classes of an equivalence relation preserved by G are equal. If H is the isotropy subgroup of x, and O_x is the orbit of x, then* $|G| = |H|\,|O_x|$.

Proof. If $(x)g_1 = y$ and $(y)g_2 = z$ then $(x)g_1 g_2 = z$. The proofs of reflexivity and symmetry are similar. By definition of transitive, G is transitive if and only if all elements of x belong to the same orbit. If there are several orbits, the relationship of belonging to the same orbit is preserved by the group. Thus G is not primitive.

Suppose G is transitive but not primitive, and $(x, y) \in R$ if and only if $((x)g, (y)g) \in R$ for an equivalence relation R, for all $g \in G$. Then $(\overline{x})g = \overline{xg}$. So for any two equivalence classes $\overline{x}$, $\overline{y}$, $(\overline{x})g = \overline{y}$ for some g. Thus, $\overline{x}, \overline{y}$ have the same cardinality.

Let H, O_x be as in the theorem. By Theorem 2.8.1, $|G| = |H|\,[G : H]$ where $[G : H]$ denotes the number of cosets of H (here we will use right cosets, but the proof is the same). By definition $O_x = \{(x)g : g \in G\}$. We have $(x)g_1 = (x)g_2$ if and only if $(x)g_1 g_2^{-1} = (x)$ if and only if $g_1 g_2^{-1} \in H$ if and only if $g_1 g_2^{-1} = h \in H$ if and only if $g_1 = hg_2$ for some $h \in H$ if and only if g_1 and g_2 belong in the same right coset of H. Thus the number of right cosets equals the size of O_x, since we have an isomorphism from cosets Hg to orbit elements $(x)g$. So $|G| = |H|\,|O_x|$. This proves the theorem. □

EXAMPLE 2.12.9. The set of permutations e, (12) (34), (13) (24), (14) (23), (1324), (1423), (12), (34) consists of all permutations which preserve the equivalence relation with partition $\{\{1,2\},\{3,4\}\}$. It follows that it is a group. It is transitive since $1 \to 1$, $1 \to 2$, $1 \to 3$, $1 \to 4$ under different elements of the group, but is not primitive. The isotropy group of 1 is e, (34). So $|G| = 8 = |H|\,|O_x| = 2 \times 4$.

The formula $|G| = |H|\,|O_x|$ is used to prove many combinatorial results, where some set related to a group must be enumerated. An example in group theory itself is the size of conjugacy classes.

PROPOSITION 2.12.4. *The order of any conjugacy class in a finite group divides the order of the group.*

Proof. We may regard G as a set of permutations of G itself by conjugation. The orbits will be the conjugacy classes. And by Theorem 2.12.3, $|O_x|$ divides $|G|$. □

EXERCISES

Level 1

1. Give representatives of all conjugacy classes in S_4.
2. For $g = (12)$ describe C_g on any permutation written on cyclic form. Try (1345) and (243) as examples. Use the last result of Section 2.7.
3. In S_4 list the number of elements in each conjugacy class.
4. What group is generated by $x = (12)$ and $y = (45)$ in the symmetric group of degree 5. What are its orbits? Find all distinct products $x^i y^j$.
5. What group is generated by $x = (123)$ and $y = (45)$ in the symmetric group of order 5. What are its orbits? Find all products $x^i y^j$. Is this group cyclic?

Level 2

1. Show a subgroup is normal if and only if it is a union of conjugacy classes.
2. Prove a group of order p^n for p prime has at least p conjugacy classes of one element. Note that the sizes of all conjugacy classes C_j are p^m for $m \geqslant 0$ and that $p^n = \Sigma\,|C_j|$. Thus

$$p^n = \sum_{|C_j|=1} |C_j| + p \sum_{\substack{|C_j|=p^r \\ r>0}} \frac{|C_j|}{p}$$

Thus

$$\sum_{|C_j|=1} |C_j|$$

is a multiple of p, being

$$p^n - p \sum_{p\,|\,|C_j|} \frac{|C_j|}{p}$$

3. Prove an element is in a conjugacy class by itself if and only if it is in the center. Therefore any p group has a nontrivial center.
4. Consider the group G of permutations of degree 5 generated by $G_1 = \{e, (12), (13), (23), (123), (132)\}$ and $G_2 = \{e, (45)\}$. Show each element of G_1 commutes with each element of G_2 and note $G_1 \cap G_2 = \{e\}$. Describe an isomorphism $S_3 \times S_2$ into G. What are the orbits of G?
5. Generalize the above exercise to represent $S_n \times S_m$ as a subgroup of S_{n+m}.
6. Prove that if G is generated by subgroups G_1, G_2 such that all elements of G_1 commute with all elements of G_2 there is a homomorphism of $G_1 \times G_2$ into G. If $G_1 \cap G_2 = \{e\}$, prove it is 1-to-1.
7. Prove a transitive permutation group of prime degree is primitive.
8. Give an automorphism of $\mathbf{Z}_2 \times \mathbf{Z}_2$, which is not the identity. Give similar automorphisms of $G \times G$ for any G, and of $G \times G \times \ldots \times G$, n factor.
9. Find all automorphisms of $\mathbf{Z}_5$. (Any automorphism is determined by $f(1)$ where 1 is a generator. Try $f(1) = 1$, $f(1) = 2$, and so on.)
10. Consider G acting on itself by $x \to xg$. Show this is transitive, but if G has a nontrivial proper subgroup H, it is not primitive (consider the right cosets Hx).

Level 3

1. Show a group is simple (has no normal subgroups) if and only if any non-identity conjugacy class generates the entire group.
2. Every permutation either preserves the sign of the product

$$D = \prod_{i<j} (x_i - x_j)$$

or reverses it if the permutation is applied to the subscripts giving

$$\prod_{i<j} (x_{p(i)} - x_{p(j)})$$

For instance for degree 2 we have

$$D = (x_1 - x_2)(x_2 - x_3)(x_1 - x_3)$$

Under (12) this goes to $(x_2 - x_1)(x_1 - x_3)(x_2 - x_3) = -D$. Prove any transposition reverses the sign of D.
3. Prove there is a homomorphism h from the symmetric group onto $\{\pm 1\}$ by sending x to ± 1 according as x preserves or reverses the sign of D. Here D is the same as in the above exercise.
4. The kernel of n is a normal subgroup of the symmetric group called the ***alternating group*** A_n. Prove every product of an even number of transpositions is in A_n but no product of an odd number of transpositions is.
5. Prove that A_5 is simple by showing each of its 3 non-identity conjugacy classes generates it. It is the smallest noncommutative simple group.

6. Let W be the subgroup of S_{nm} such that W preserves the partition $\{1, 2, \ldots, m\}$, $\{m+1, m+2, \ldots, 2m\}$, $\ldots$, $\{nm - m + 1, \ldots, nm\}$. Show W is transitive but not primitive. Show W has a normal subgroup W_0 isomorphic to $S_m \times S_m \times \ldots \times S_m$ which fixes each set of the partition. Show W/W_0 is isomorphic to S_n. What is the order of W? It is called the ***wreath product*** of S_n and S_m.

2.13 SYMMETRIES

A symmetry of some object is a 1-to-1 onto mapping from the object to itself which preserves some of the structure of the object. One of the most characteristic mathematical methods is to look for the symmetries in a problem.

EXAMPLE 2.13.1. The operation $x \to x^c$ establishes a symmetry from the algebra of sets to itself which reverses the operation $\cap$, and $\cup$. Thus any theorem which is true for $\cap$ is also true for $\cup$. For instance since $A \cap (B \cup C) = (A \cap B) \cup (A \cap C)$ we have also $A \cup (B \cap C) = (A \cup B) \cap (A \cup C)$.

EXAMPLE 2.13.2. The equation $x^4 + x^2 + 1$ is unchanged if we replace x by $-x$. Thus the negative of any root is also a root.

EXAMPLE 2.13.3. For any group G there is a 1-to-1 onto mapping G to G which sends x to x^{-1}. Since $(xy)^{-1} = y^{-1}x^{-1}$ this mapping reverses the order of products, so that right cosets become left cosets and left cosets become right cosets. Thus for any property of left cosets there is a similar property of right cosets.

EXAMPLE 2.13.4. Conjugation $x + iy \to x - iy$ is a symmetry of the complex number system.

EXAMPLE 2.13.5. Suppose we want to show that the edges of the figure below (called a ***Petersen graph***) cannot be 3-colored so that edges of all 3 colors

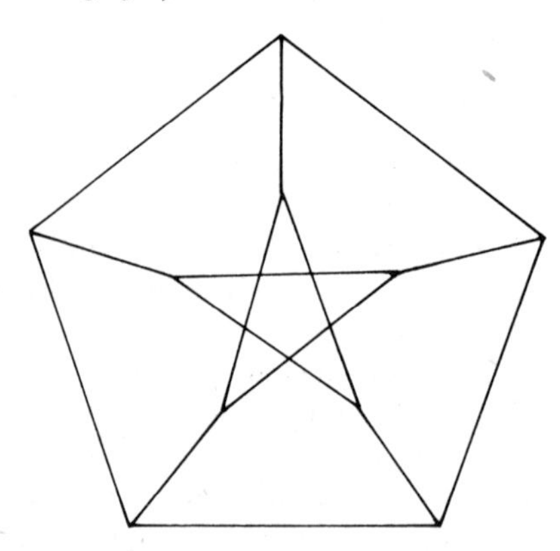

meet at each vertex. We can reason as follows. Note that the entire figure has a pentagonal symmetry. Any 3-coloration of a pentagon uses all 3 colors since a pentagon cannot be 2-colored. There are symmetries of such a graph coloring problem which permute the colors in any way. That is, if we have one coloring we may obtain another by recoloring all 1 edges as 2, all 2 edges as 3, and all 3 edges as 1, for instance. In a pentagon not all 3 colors can appear on 2 edges. So assume color 3 occurs only on 1 edge. No color may appear on 3 edges of a pentagon because some two of the three will meet. So each of colors 1, 2 appear on 2 edges. If we remove the edge colored 3 the other colors must alternate. So we may say that any edge coloring on a pentagon, up to symmetry is

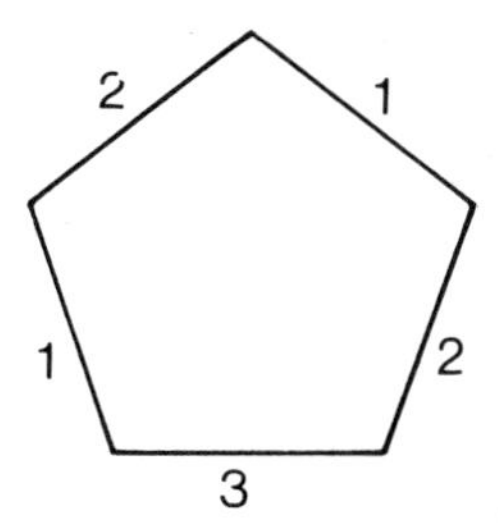

So we may suppose that the outer edges of the Petersen graph are colored in this way. The edges between the outer and inner edges are determined by this. We have

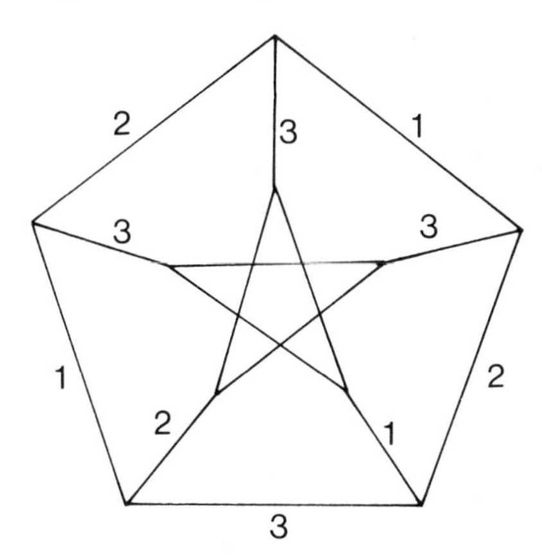

Evidently now no inside edges can be colored 3. But the inner 5 edges form a pentagon. So one of them must be colored 3. This proves the impossibility of 3-coloring a Petersen graph. This problem is related to the 4-color problem, see Gardner (1976).

The 4-color theorem asserts that any map in the plane can be colored with 4 colors in such a way that no two regions of the same color have an extended common boundary larger than a point. This can be reduced to the case of maps such that no more than three regions meet at a point, called ***trivalent maps.*** A trivalent map can be 4-colored if and only if its edges can be 3-colored. The Petersen graph (which, however, is not planar) is the smallest trivalent graph which cannot be 3-colored and is 'bridgeless'. Tutte has conjectured that *every bridgeless trivalent graph which cannot be 3-colored contains a topological copy of the Petersen graph.*

EXAMPLE 2.13.6. The set of $n \times n$ matrices has the symmetry $X \rightarrow X^{\mathrm{T}}$, the transpose of X. This means that for any theorem about the columns of a matrix there is a corresponding theorem about the rows of a matrix.

EXAMPLE 2.13.7. For any partial order there is another partial order obtained by replacing $\leqslant$ by $\geqslant$. This means for any theorem about $\geqslant$ there is a corresponding theorem about $\leqslant$.

Symmetries such as in Examples 2.13.1, 2.13.3, 2.13.4, 2.13.6, 2.13.7 are called ***dualities***.

Symmetry was used by Felix Klein to classify different kinds of geometry. Euclidean geometry is based on what is called the ***group of rigid motions***, which are like drawing a figure on a piece of paper and moving it around on top of a plane. Any two congruent figures can be made to coincide by a rigid motion applied to one of them. ***Euclidean geometry*** is the study of properties like distance and angle which are invariant under rigid motions, that is, they do not change.

Projective geometry studies properties unchanged under a large group of transformations, which include the transformations which occur when a shadow falls on a plane inclined at an angle or a person sees something tilted at an angle to his field of vision. Here angles and distances can change according to the position of the observer, and other invariant quantities, such as cross-ratios, are studied. A circle may be changed into an ellipse or even a hyperbola. Projective geometry gives new results in ordinary geometry, such as Pascal's theorem that the extended sides of a hexagon inscribed in a conic section meet by pairs in three collinear points.

EXAMPLE 2.13.8. Symmetry groups are used in the classification of elementary particles. Sometimes a group of particles can be considered as separate states of a single particle. These states will be solutions of an equation having a certain symmetry, derived from a symmetry of a force law. If there is also a weaker force which does not have the symmetry, solutions will come in sets that are almost but not quite symmetric.

The following analogy may convey a little of what is involved. Suppose we have an algebraic polynomial equation of degree 4 and a symmetry $x \to -x$ (actually there are partial differential equations and infinite, continuous symmetry groups). Then the equation has a form such as

$$x^4 - ax^2 + b = 0.$$

Roots occur in symmetrical pairs $\pm r$. Now suppose a small nonsymmetric term is added

$$x^4 - ax^2 + 0.001ax + b = 0.$$

Then the roots will still probably come in pairs which are almost but not quite equal in absolute value. This is analogous to particles having nearly but not quite the same mass, like the proton and neutron.

For the remainder of this section we will be concerned with geometrical symmetries. Let E^n denote n-dimensional Euclidean space. That is, points of E^n are n-tuples $(x_1, x_2, \ldots, x_n)$ of real numbers, distance is given by the formula

$$\sqrt{(x_1 - y_1)^2 + (x_2 - y_2)^2 + \ldots + (x_n - y_n)^2}$$

and the angle between two vectors v, w is defined by

$$\cos \Theta = \frac{v \cdot w}{|v|\,|w|}$$

where the angle Θ is chosen to lie between 0 and π.

DEFINITION 2.13.1. An *isometry* of E^n is a one-to-one, onto function $f: E^n \to E^n$ such that for all points x, y the distance $d(x, y)$ equals the distance $d(f(x), f(y))$. Two subsets S_1, S_2 of E^n are congruent if and only if $f(S_1) = S_2$ for some isometry f.

EXAMPLE 2.13.9. For any $(a_1, a_2, \ldots, a_n)$, the mapping $(x_1, x_2, \ldots, x_n) \to (x_1 + a_1, x_2 + a_2, \ldots, x_n + a_n)$ is an isometry called a *translation*. A translation moves the plane a distance $\sqrt{a_1^2 + a_2^2 + \ldots + a_n^2}$ in the direction of the vector $(a_1, a_2, \ldots, a_n)$. It does not alter the directions of lines.

EXAMPLE 2.13.10. The transformation $x \to Ax$ for a matrix A will be an isometry if and only if $A^T = A^{-1}$. Such matrices are called ***orthogonal***.

DEFINITION 2.13.2. In E^3 a *reflection* in the plane H is the function such that $f(x) = y$ where H is the perpendicular bisector of the segment from x to y, and $f(x) = x$ if $x \in H$. Reflections through a line in E^2 are defined similarly. A reflection through a point p is defined by $f(x) = y$ where p is the midpoint of the segment from x to y.

All reflections are isometries.

EXAMPLE 2.13.11. The reflection in the x, y plane sends (x, y, z) to $(x, y, -z)$.

A rotation in E^3 will be considered as any isometry which leaves a certain line (the axis) pointwise fixed, and does not fix any other point.

THEOREM 2.13.1. *Every isometry f can be written as tg where t is a translation and g is an isometry which fixes the point $(0, 0, \ldots, 0)$.*

Every isometry which fixes $(0, 0, \ldots, 0)$ is multiplication of every vector by a unique orthogonal matrix. Isometries form a group in which translations are a normal subgroup.

Proof. Let t be the translation which takes $(0, 0, \ldots, 0)$ to $f^{-1}(0, 0, \ldots, 0)$. Then $tf = g$ fixes $(0, 0, \ldots, 0)$.

An isometry which fixes $(0, 0, \ldots, 0)$ will preserve distance by definition. It will preserve straight lines since a straight line is the shortest distance between two points. It will preserve the lengths of the sides of all triangles, so will also preserve angles. It will thus preserve parallelism of lines. So it will send a sum of two vectors to a sum of vectors. Thus it will satisfy $f(v + w) = f(v) + f(w)$. This implies $f(kw) = kf(w)$ for all rational numbers k. By a continuity argument, $f(kw) = kf(w)$ for all real numbers k. Thus f is a linear transformation and can be represented by a matrix. This matrix will be orthogonal.

It is straightforward to show that if f, g are isometries so are fg and f^{-1}. So the isometries form a group. The composition of the translations $(x_1, x_2, \ldots, x_n) \to (x_1 + a_1, x_2 + a_2, \ldots, x_n + a_n)$ and $(x_1, x_2, \ldots, x_n) \to (x_1 + b_1, x_2 + b_2, \ldots, x_n + b_n)$ is $(x_1, x_2, \ldots, x_n) \to (x_1 + a_1 + b_1, x_2 + a_2 + b_2, \ldots, x_n + a_n + b_n)$. The inverse of the former is $(x_1, x_2, \ldots, x_n) \to (x_1 - a_1, x_2 - a_2, \ldots, x_n - a_n)$. Let t_1 be a translation $x \to x + a_1$ in vector notation and $f = t_2 g$ an isometry, where g is represented by an orthogonal matrix A. Then $(x)ft_1f^{-1} = (x)t_2gt_1g^{-1}t_2^{-1} = (x + a_2)At_1A^{-1}t_2^{-1}$, a translation, where $(x)t_2 = x + a_2$. Thus translations form a normal subgroup. This proves the theorem. □

Later we will take up matrices in more detail, but for the present we will state some facts without proof.

DEFINITION 2.13.3. The *eigenvalues* of an $n \times n$ matrix A are the roots of the equation $\det(zI - A) = 0$, a polynomial of degree n.

EXAMPLE 2.13.12. The eigenvalues of a 2×2 matrix A are the roots of

$$\det \begin{bmatrix} z - a_{11} & -a_{12} \\ -a_{21} & z - a_{22} \end{bmatrix} = 0$$

or $(z - a_{11})(z - a_{22}) - a_{12}a_{21} = 0$.

DEFINITION 2.13.4. An *eigenvector* corresponding to an eigenvalue z is a vector v such that $v(zI - A) = 0$, or $zv = zIv = vA$.

EXAMPLE 2.13.13. For a diagonal matrix the vector v having a 1 in place i and 0s elsewhere is an eigenvector. In fact $vA = a_{ii}v$.

Orthogonal 3×3 real matrices have the following forms. All eigenvalues of an orthogonal matrix have absolute value 1, and satisfy a cubic equation with real coefficients. Thus either all three are real (and must be ± 1) or two are complex and one is real.

Eigenvalues	Operation
$1, 1, 1$	Identity
$-1, 1, 1$	Reflection in plane of last two eigenvectors
$-1, -1, 1$	180° rotation about axis of last eigenvector
$-1, -1, -1$	$-I$, reflection through origin
$1, z, \overline{z}$	Rotation about axis of first eigenvector
$-1, z, \overline{z}$	Combined rotation about axis of first eigenvector and reflection in the plane of last two eigenvectors

In general every orthogonal matrix of determinant 1 will be a rotation. Rotations of a figure can be physically realized by moving the object about, whereas reflections cannot.

Every 2×2 orthogonal matrix has one of these two forms, where $x^2 + y^2 = 1$:

$$\begin{bmatrix} x & y \\ -y & x \end{bmatrix}, \begin{bmatrix} x & y \\ y & -x \end{bmatrix}$$

The former, of determinant 1, will be a rotation, and the latter, of determinant -1, will be a reflection.

DEFINITION 2.13.5. If S is a subset of E^n, the *group of symmetries* of S is the group of all functions $f: S \to S$ such that there exists an isometry g with $f(s) = g(s)$ for all $s \in S$.

EXAMPLE 2.13.14. The group of all symmetries of the line segment $[-1, 1]$ is the group consisting of the identity function and the function $f(x) = -x$.

THEOREM 2.13.2. *The group of symmetries of a regular polygon of n sides is a group of order 2n having a normal subgroup isomorphic to* $\mathbf{Z}_n$. *Its elements can be represented by* $e, z, z^2, \ldots, z^{n-1}, y, yz, \ldots, yz^{n-1}$ *where products can be computed by the rules* $z^n = e$, $y^2 = e$, $z^j y = yz^{-j}$.

Proof. Assume the polygon has its center at the origin, that the distance from its center to a vertex is 1, and that a vertex v_1 is located at the point $(1, 0)$. All isometries of E^n preserve convex combinations since both multiplication by isometries and translations do. The center is the vector $\frac{1}{n}(v_1 + v_2 + \ldots + v_n)$ if v_n are the vertices, so the center is preserved. Thus all isometries are rotations and reflections about the origin. Any rotation is determined by the position of v_1. Thus there are exactly n rotations, which send v_1 to $v_1, v_2, \ldots, v_n$. Let z be the rotation sending v_1 to v_2. Then the others are $z, z^2, \ldots, z^{n-1}$ and $z^n = e$.

The rotations are the isometries in the kernel of the homomorphism $\det(A)$ to the group $\{\pm 1\}$. Thus they form a normal subgroup K. Let y be the reflection $(x, y) \rightarrow (x, -y)$. Then y is a symmetry of the polygon. Any reflection would satisfy the rule $y^2 = e$. And yzy sends v_1 to v_n. Its determinant is $(-1)(1)(-1)$ so it is a rotation. So $yzy = z^{-1}$. By the properties of conjugation $yz^jy = yz^jy^{-1} = z^{-j}$. So $yz^j = z^{-j}y$ Finally we must observe that no other isometries exist. The subgroup K has two cosets by Theorem 2.5.1 since the image under $\det(A)$ is the subgroup $\{\pm 1\}$. Each coset has n elements, so the group has order precisely $2n$. This completes the proof. □

The group described by this theorem is an important group known as the *dihedral group*. Its elements can be represented in matrix form as follows:

$$z^j = \begin{bmatrix} \cos \frac{2j}{n} & \sin \frac{2j}{n} \\ -\sin \frac{2j}{n} & \cos \frac{2j}{n} \end{bmatrix}, \quad y = \begin{bmatrix} 1 & 0 \\ 0 & -1 \end{bmatrix}$$

EXERCISES

Level 1

1. What is a symmetry of the function $\sin x$?
2. The equation $x^4 + x^3 + x^2 + x + 1 = 0$ can be written as $x^2 + x + 1 + x^{-1} + x^{-2} = 0$. Show that replacing x by x^{-1} is a symmetry of this.
3. Use this symmetry to reduce $x^2 + x + 1 + x^{-1} + x^{-2}$ to a quadratic in $z = x + x^{-1}$. Note that $z^2 + x^2 + x^{-2} + 2$. Write the function as a quadratic expression $z^2 + az + b$. Find its roots in z.
4. A nonsquare rectangle has a symmetry group consisting of e, one rotation and two reflections. Describe the rotation and reflections.
5. Draw a 3-coloring of the edges of a cube. How many different 3-colorings exist? How many coincide under a symmetry of the cube or a permutation of the colors?
6. Write out the multiplication table of the dihedral group of order 10, using the description in Theorem 2.13.2.

Level 2

1. Color the faces of a cube with as few colors as possible so that no two faces of the same color have an edge in common.
2. Every finite partial order has a maximum element. What is the dual to this theorem?
3. Write out the Kronecker product of this matrix with itself (which will be a 4×4 matrix) $(b_{2i+k-2,2j+m-2}) = (a_{ij}a_{km})$, where i, j, k, m range over 1, 2. To find b_{rs} put in all values for i, j, k, m successively. If $i = k = j = m = 1$, $b_{2+1-2,2+1-2} = a_{11}a_{11}$ so $b_{11} = a_{11}^2 = 1^2 = 1$.

$$A = \begin{bmatrix} 1 & 2 \\ 3 & 5 \end{bmatrix}$$

   Show the Kronecker product of its square is the square of its Kronecker product.
4. Prove the Kronecker product $\boxtimes$ is multiplicative in that $(A \boxtimes B)(C \boxtimes D) = (AB \boxtimes CD)$, where $(A \boxtimes B)_{ni+k-n,nj+m-n} = a_{ij}b_{km}$, B is $n \times n$.
5. Prove that any two translations commute. Show the group of translations in E^n is isomorphic to $\mathbf{R} \times \mathbf{R} \times \ldots \times \mathbf{R}$ where $\mathbf{R}$ is the additive real numbers.
6. Show that the group of translations and rotations in the plane is isomorphic to the group of transformations $az + b$ on complex numbers, where a, b are complex numbers and $|a| = 1$. What operation corresponds to reflection in the x-axis?
7. The only rotations that can occur in symmetries of crystals are multiples of 60° and 90°. This can be shown as follows. The atoms in a plane section form a regular structure of points called a ***lattice***. If one atom is fixed as $(0, 0)$ all atoms will be a set L of points (x, y) such that if $(x, y) \in L$ and $(w, z) \in L$ then $(x + w, y + z) \in L$ and $(-x, -y) \in L$. That is, it will be a subgroup of $\mathbf{R} \times \mathbf{R}$. In order for atoms to be a finite distance apart in 2-dimensions the lattice must have 2 generators, z, w. If A is the matrix of the rotation, $Az = n_1 z + n_2 w$, $Aw = n_3 z + n_4 w$, $n_1, n_2, n_3, n_4 \in \mathbf{Z}$. Thus the ***trace*** (sum of diagonal entries in any basis) is $n_1 + n_4$, an integer. The trace is independent of basis. For a standard basis a rotation will be represented as

$$\begin{bmatrix} \cos\theta & \sin\theta \\ -\sin\theta & \cos\theta \end{bmatrix}$$

   with trace $2\cos\theta$. Thus $2\cos\theta = n_1 + n_4$, an integer. Show this implies θ is a multiple of 60° or 90°. Give examples of lattices (or regular tilings) with 60° and 90° symmetry.

Level 3

1. The linear system

$$\begin{cases} y' = ay + bz \\ z' = bz + ay \end{cases}$$

is invariant under interchanging y and z. Show that it splits into separate equations on $u = y + z$ and $v = y - z$.

2. Let A be a matrix and G a group of matrix symmetries of A, that is, $XA = AX$ for $X \in G$. Then show that if y is any linear combination of the matrices $X \in G$, show $Ay = yA$. What generally happens is that a collection $y\langle i\rangle$ of primitive idempotents can be found satisfying $\Sigma\, y\langle i\rangle = I$, $y\langle i\rangle y\langle j\rangle = 0$ for $i \neq j$, $y^2\langle i\rangle = y\langle i\rangle$. Show that any vector x is the summation of $x\langle i\rangle = y\langle i\rangle x$. This represents the group V of all vectors x as the direct product set $(x\langle 1\rangle, x\langle 2\rangle, \ldots, x\langle n\rangle)$ of $V\langle i\rangle = Y\langle i\rangle V$. Show that $Ax = 0$ if and only if $Ax\langle i\rangle = 0$ for each i. This is the basic reason for the physics example.

3. Show that the group of isometries in E^n can be represented as a group of $(n+1)$-dimensional matrices with block form

$$\left[\begin{array}{c|c} B & 0 \\ \hline A & I \end{array}\right]$$

where B is an orthogonal matrix acting on row vectors and $A = (a_1, a_2, \ldots, a_n)$ represents the translation.

4. Prove that $|Ax|^2 = x^2$ for all vectors x if and only if $A^{\mathrm{T}} = A^{-1}$. Since $x \cdot y = \frac{1}{2}(|x + y|^2 - |x|^2 - |y|^2)$, $|Ax|^2 = x^2$ is equivalent to $Ax \cdot Ay = x \cdot y$ for all x, y, or since A must be nonsingular $Ax \cdot z = x \cdot A^{-1}z$ where $z = Ay$. Prove $Ax \cdot z = x \cdot A^{\mathrm{T}}z$ and that $Ax \cdot z = Bx \cdot z$ for all x, z implies $A = B$.

5. Show the orthogonal matrices are closed under multiplication. Show a matrix is orthogonal if and only if its rows A_{i*} satisfy $|A_{i*}|^2 = 1$ and $A_{i*} \cdot A_{j*} = 0$ for $i \neq j$. Here A_{i*} denotes the ith row of A. Note that $A^{\mathrm{T}} = A^{-1}$ is equivalent to $AA^{\mathrm{T}} = I$.

6. Compute the orders of the symmetry groups of a tetrahedron, an octahedron, and an icosahedron. These groups are transitive, so the number of vertices can be multiplied by group of symmetries which fix a given vertex. There will be the same number of reflections and rotations.

7. Show the group of symmetries of a tetrahedron is isomorphic to the symmetric group on its vertices.

8. Show the groups of rotations of a cube is isomorphic to the symmetric group on its diagonals.

9. Show the group of rotations of an icosahedron is isomorphic to the alternating group.

2.14 POLYA ENUMERATION

Polya enumeration is concerned with counting the number of patterns of a certain type where two symmetrical patterns are considered the same.

EXAMPLE 2.14.1. Consider all labellings of the vertices of a square with $\{0, 1\}$. There exist $2^4 = 16$ such labellings since each vertex can be labelled 2 ways no matter how previous vertices have been labelled. But if we consider symmetry there are precisely 6 distinct labellings:

$$\begin{bmatrix} 0 & 0 \\ 0 & 0 \end{bmatrix}, \begin{bmatrix} 1 & 0 \\ 0 & 0 \end{bmatrix}, \begin{bmatrix} 1 & 0 \\ 1 & 0 \end{bmatrix}, \begin{bmatrix} 1 & 0 \\ 0 & 1 \end{bmatrix}$$

$$\begin{bmatrix} 1 & 1 \\ 1 & 0 \end{bmatrix}, \begin{bmatrix} 1 & 1 \\ 1 & 1 \end{bmatrix}$$

G. Polya's formula follows from a lemma of W. Burnside. We state this as in W. Gilbert (1976).

LEMMA 2.14.1. (Burnside.). *Let G be a group acting on a finite set X. For $g \in G$ let f_g be the number of elements fixed by g. Then the number of orbits of G is $|G|^{-1} \Sigma f_g$.*

Proof. Let $S = \{(x, g): g(x) = x\}$. For each element $g \in G$ we have f_g members of S. So $|S| = \Sigma f_g$. On the other hand for a fixed X the number of elements g such that $g(x) = x$ equals the isotropy group of x. This has size $|G|\,|O_x|^{-1}$ by Theorem 2.12.3. So any orbit contributes in all

$$|O_x| \frac{|G|}{|O_x|}$$

elements. If N is the number of orbits, $|S| = N|G| = \Sigma f_g$. □

In the following we consider a set X acted on by a group G. Each element of X is labelled by a member of a set L of labels. (This is equivalent to a function $f: X \to L$.) Two labellings, or functions, are equivalent if they differ by an operation of G. The problem is to find the number of nonequivalent labellings.

EXAMPLE 2.14.2. Suppose G is the symmetric group. Then two labellings are equivalent if each label occurs the same number of times.

PROPOSITION 2.14.2. *The number of equivalence classes is*

$$\frac{1}{|G|} \Sigma\, |L|^{c(g)}$$

where L is the set of labels and c(g) is the number of cycles of G.

Proof. We need to show $f_g = |L|^{c(g)}$. For fixed g choose cycle representations $x_1, x_2, \ldots, x_{c(g)}$. To show that a labelling is invariant means precisely that all members of the same cycle receive the same labelling. These invariant labellings are in 1-1 correspondence with labellings of $x_1, x_2, \ldots, x_{c(g)}$. Each x_i can be labelled $|L|$ ways regardless of how previous ones are labelled. Thus there are $|L|^{c(g)}$ ways to label $x_1, x_2, \ldots, x_{c(g)}$. □

Labellings can be classified more completely by Polya's general result.

Associate a variable α_i to each element of L. Let $P_G(x_1, x_2, \ldots, x_k)$ denote the polynomial

$$\frac{1}{|G|} \sum_{g \in G} x_1^{c_1(g)} x_2^{c_2(g)} \ldots x_k^{c_k(g)}$$

where $c_i(g)$ is the number of cycles of g of length i, and k is the maximum length of any cycle.

EXAMPLE 2.14.3. For G a cyclic group of prime order contained in the symmetric group of degree p there will exist the identity transformation with p 1 cycles and $p - 1$ nonidentity elements with 1 p cycle. So

$$P_G(x_1, x_2, \ldots, x_p) = \tfrac{1}{p}(x_1{}^p + (p-1)x_p{}^1)$$

THEOREM 2.14.3. (Redfield and Polya.) *The number of labellings in which α_i occurs precisely n_i times is the coefficient of $\alpha_1{}^{n_1}\alpha_2{}^{n_2} \ldots \alpha_m{}^{n_m}$ in*

$$P_G(\alpha_1 + \alpha_2 + \ldots + \alpha_m, \alpha_1{}^2 + \alpha_2{}^2 + \ldots + \alpha_m{}^2, \ldots, \alpha_1{}^k + \alpha_2{}^k + \ldots + \alpha_m{}^k)$$

Proof. We apply Lemma 2.14.1 to the set S of labellings in which α_i occurs exactly n_i times. By the lemma the number of equivalence classes equals

$$\frac{1}{|G|} \sum_g |s \in S : (s)g = s|$$

Suppose g has $c_i(g)$ orbits of length g. Then it suffices to show

$$|s \in G : (s)g = s|$$

equals the coefficient of $\alpha_1{}^{n_1}\alpha_2{}^{n_2} \ldots \alpha_m{}^{n_m}$ in

$$(\alpha_1 + \alpha_2 + \ldots + \alpha_m)^{c_1(g)} \ldots (\alpha_1{}^k + \alpha_2{}^k + \ldots + \alpha_m{}^k)^{c_k(g)}.$$

Choose representatives x_{ij} of all j-cycles. Let $S_j = \{x_{ij}\}$. The the labellings of X invariant under g are the same as labellings of these representatives. The condition on the number of labels is equivalent to saying that the number of occurrences of α_i in S_i plus twice the number of occurrences in S_2, and so on, equals n_i. Note that $c_i(g) = |S_i|$. Thus the number of choices mentioned equals the number of ways to choose a term from each of the following

$$c_1(g) \text{ times}$$

$$(\alpha_1 + \alpha_2 + \ldots + \alpha_m), \ldots, (\alpha_1 + \alpha_2 + \ldots + \alpha_m), \ldots,$$

$$c_k(g) \text{ times}$$

$$(\alpha_1{}^k + \alpha_2{}^k + \ldots + \alpha_m{}^k), \ldots, (\alpha_1{}^k + \alpha_2{}^k + \ldots + \alpha_m{}^k)$$

so that the total exponent of α_i is n_i. But this equals the coefficient of

$$\alpha_1{}^{n_1} \alpha_2{}^{n_2} \ldots \alpha_m{}^{n_m}$$

in the product. □

EXAMPLE 2.14.4. Consider labellings of a square with $0, +1, -1$. The symmetry group of a square consists of the identity (8, 1-cycles), two 90° rotations (1 4-cycle), a 180° rotation (2 2-cycles), two reflections on a vertical or horizontal axis (2 2-cycles), and two reflections on a diagonal axis (1 2-cycle, 2 1-cycles).

So the polynomial P_G is

$$\tfrac{1}{8}(x_1{}^4 + 2x_4 + 3x_2{}^2 + 2x_2 x_1^2)$$

The various labellings are then given by

$$\tfrac{1}{8}((\alpha_1 + \alpha_0 + \alpha_{-1})^4 + 2(\alpha_1^4 + \alpha_0^4 + \alpha_{-1}^4) + 3(\alpha_1^2 + \alpha_0^2 + \alpha_{-1}^2)^2$$
$$+ 2(\alpha_1^2 + \alpha_0^2 + \alpha_{-1}^2)(\alpha_1 + \alpha_0 + \alpha_{-1})^2)$$

For instance the coefficient of $\alpha_1{}^2 \alpha_0 \alpha_{-1}$ is

$$\tfrac{1}{8}(12 + 2(2)) = 2.$$

The two labellings are

$$\begin{bmatrix} 1 & 1 \\ 0 & -1 \end{bmatrix}, \begin{bmatrix} 1 & 0 \\ -1 & 1 \end{bmatrix}$$

EXERCISES

Level 1

1. Work out the labellings of a square by $\{0, 1\}$ using Polya's method. Imitate the last example but use only α_0, α_1.

2. Write out all distinct labellings of a pentagon with 0, 1.
3. Work out this problem by Polya's method where G, the dihedral group, has the identity with 5 1-cycles, 4 rotations with 1 5-cycle, and 5 reflections with 2 2-cycles and 1 1-cycle.
4. Write the labellings of a hexagon with 0, 1 so that there are 3 zeros and 3 ones.
5. Compute the number of ways to label the vertices of a square with 4 labels if all 4 labels appear. Check by writing out the diagrams.

Level 2

1. In switching circuits designers are concerned with functions on n Boolean variables. Thus we have a set of 2^n elements (all values of $x_1, x_2, \ldots, x_n$) acted on by the permutations of $x_1, x_2, \ldots, x_n$. These are labelled with 0, 1. Use Polya's theory to compute the number of nonequivalent functions of this kind for $n = 2$. Example: the following are distinct.

(1)

x_1	x_2	f
0	0	0
0	1	0
1	0	0
1	1	0

(2)

x_1	x_2	f
0	0	1
0	1	0
1	0	0
1	1	0

(3)

x_1	x_2	f
0	0	1
0	1	1
1	0	0
1	1	0

The group has two elements, the identity and (12). The latter acts with a 2-cycle and 2 1-cycles.

2. Consider switching circuits with a larger group generated by both permutation and complementation of input variables. For the group has $n!\,2^n$ elements for n variables. For $n = 2$ we have

Number of elements	Cycles of order 1	2	3	4
1	4	0	0	0
2	1	0	0	1
3	0	2	0	0
2	2	1	0	0

(In fact it is the symmetry group of a square.)

3. Write out the functions for Exercise 2.
4. Do Exercise 2 for three variables. The group will be the symmetry group of a cube, of order 48.
5. Find a general formula for the number of 'necklaces' of black and white beads. This is the same problem as labelling a regular n-gon with $\{0, 1\}$. The

group is the dihedral group of order $2n$ and has the following elements for n odd (consider only this case here), where d is any divisor of n: $\phi(d)$ rotations having n/d d-cycles and n reflections having 1 1-cycle and $\frac{n-1}{2}$ 2-cycles. Here $\phi(n)$ is ***Euler's function*** $n\,\Pi\,(1-\frac{1}{p})$ where p ranges over prime divisors of n. Check the formula for $n = 5, 7$.

Level 3

1. Solve the switching circuits problem for $n = 4$ with complementation (see Exercise 2 of Level 2).
2. Solve the necklace problem for n even (see Exercise 5 of Level 2).
3. How many distinct molecules can have this configuration

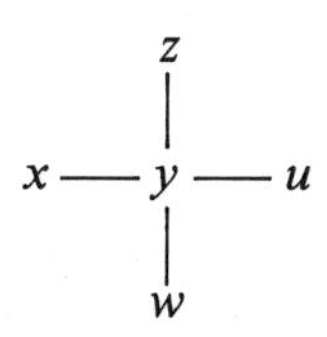

where each letter represents one of three given atoms?
4. Compute the number of distinct cyclic strings of 0, 1 of length n. Two are equivalent if they can be made to coincide by taking the last element. This is like the necklace problem except there are no reflections.
5. Compute the number of ways to color the vertices of a cube with 4 colors such that each color occurs twice.

2.15 KINSHIP SYSTEMS

Certain tribal societies are divided into subsets which will here be called *clans*, though they are not hereditary, and there are strict rules about which clans a person of a given clan may marry. Probably such rules help prevent incest and help unite the tribe. These systems may have a mathematical structure. A number of workers have attempted to state this structure precisely. Here we will follow the treatment of H. C. White (1963), whose axioms can be stated as follows.

KS1. *The entire population is partitioned into n nonempty clans.*
KS2. *There is a permanent rule fixing the single clan among whose women the men of a given clan must find wives.*
KS3. *Men from two different clans cannot marry women of the same clan.*
KS4. *All children of a couple are assigned to a single clan uniquely determined by the clans of their mother and father.*
KS5. *Children whose fathers are in different clans must be in different clans.*
KS6. *A man can never marry a woman of his own clan.*

KS7. *Every person has some relative by marriage and descent in each other clan.*

KS8. *Whether two people who are related by marriage and descent are in the same clan depends only on their relationship, not on which clan either belongs to.*

Let the clans be numbered $1, 2, \dots, n$. Let w be the function such that $w(i)$ is the clan of the wife of a person in clan i. By 1, 2 there is such a function. By (KS3) it is a permutation. Let d be the function such that $d(i)$ is the clan of the children of a man in clan i and a woman, necessarily in clan $w(i)$. By (KS4), d exists, and by (KS5) it is a permutation. (KS7) states that the group G of permutations generated by w, d is transitive. (KS8) states that if for some i and some element g of the group $G, (i)g = i$ then $(j)g = j$ for all j. (KS7) now implies that w is not the identity. Moreover any permutations w, d satisfying these conditions give a possible kinship system.

We will now relate this to a theorem in group theory.

DEFINITION 2.15.1. Let G be a group of n elements, numbered $1, 2, \dots, n$. Let r be the function from G to the symmetric group of degree n such that $r(g)$ is the function f_g, where $(x)f_g = xg$. Then r is called the *regular representation* of G.

The regular representation is a particular way of associating a permutation with each element of a group.

EXAMPLE 2.15.1. Let the elements be labelled as in Example 2.8.3. Then $r(x)$ is the permutation

$$\begin{pmatrix} e & a & b & x & y & z \\ x & y & z & e & a & b \end{pmatrix}$$

from the x column of the table.

Thus the regular representation represents any group as a group of permutations. The permutation is that given by right multiplication.

THEOREM 2.15.1. (Cayley.) *The regular representation is an isomorphism from G into the symmetric group of degrees n. Thus any group is isomorphic to a subgroup of a symmetric group.*

Proof. $(x)f_{gh} = xgh = (xg)h = ((x)f_g)f_h$. Thus $r(gh) = r(g)r(h)$. So r is a homomorphism. For any x, if $r(g) = r(h)$ then $xg = xh$ so $g = h$. Therefore r is one-to-one. So G is isomorphic to the image of r. This proves the theorem. □

THEOREM 2.15.2. *A group G of permutations satisfies the conditions (1) G is transitive, (2) if $(i)g = i$ for some i then $(j)g = j$ for all j, if and only if G is the regular representation of itself.*

Proof. We first show the regular representation satisfies these conditions. For any x, y let $g = x^{-1}y$. Then $(x)f_g = xg = xx^{-1}y = y$. Thus the representation is transitive. Suppose $(x)f_g = x$. Then $xg = x$ so $g = e$.

Now suppose G satisfies (1), (2). We number the elements of G as follows: for any element g, number it $(1)g$. By condition (2) these numbers are distinct and by Condition (1) they include all of $1, 2, \ldots, n$. Now $r(h)$ is the permutation which takes $(1)g$ to $(1)gh$. But that is precisely the effect of the permutation h. So $r(h) = h$. So G is the regular representation of itself, if its elements are propertly numbered. This proves the theorem. □

COROLLARY 2.15.3. *There is a 1-to-1 correspondence between kinship structures of n clans and triples* (G, w, d) *where G is a group of order n generated by two elements, w, d are generators of G, and* $w \neq e$.

In practice as many as 16 clans may effectively occur.

EXAMPLE 2.15.2. The Kariera tribe of Australia has been associated with this marriage system although some regard it as an oversimplification.

Horizontal lines denote the tribe of a husband or wife, vertical ones those of a parent or child.

EXAMPLE 2.15.3. The Arunta tribe of Australia has been associated with this marriage system, where horizontal and vertical lines denote spouse and lines at a 45° angle, parent or child.

EXERCISES

Level 1

1. Graph the kinship structure corresponding to $\{\mathbf{Z}_4, 1, 3\}$. Vertices are clans and edges are labelled d or w from one clan to another.
2. Graph all kinship structures with group $\mathbf{Z}_2$.

3. Do the same for $\mathbf{Z}_3$.
4. Write out the complete regular representation of the symmetric group of degree 3.
5. Write out the regular representation of the cyclic group $\mathbf{Z}_5$.

Level 2

1. What is the condition that the clans of children be distinct from those of their mother?
2. What is the condition that mothers, fathers, and children belong to three different clans?
3. Find the smallest examples of the Exercise 1.
4. Find the smallest examples of the Exercise 2.
5. Work out Example 2.15.2. as a regular representation.
6. Do the same for Example 2.15.3.

Level 3

1. What is the condition that first-cousin marriages be impossible? There are four cases: the mothers are sisters, the fathers are brothers, the father of one is the brother of the mother of the other. Show the first two cannot happen.
2. Find the smallest example for which the first exercise of this level holds.
3. Prove a group of order p^2 is commutative. By an earlier exercise, the center must be at least $\mathbf{Z}_\mathbf{p}$.
4. Show for any subgroup H of a group G, G acts as permutations of the cosets of H by $(Hx)g = H(xg)$.
5. Show every transitive permutation representation arises as in the above exercise where H is the isotropy group of an arbitrary chosen point.

2.16 LATTICES OF SUBGROUPS

In this section we briefly describe some results about lattices of normal subgroups of a group. This theory also applies to modules, vector spaces, and ideals in rings.

Recall a lattice is a partially ordered set in which every pair of elements have a ***least upper bound*** (***join***, denoted $\vee$) and a ***greatest lower bound*** (***meet***, denoted $\wedge$). Any linearly ordered set, the subsets of any set, any Cartesian product of lattices forms a lattice. For intersection and union of sets (and for any linearly ordered set), the distributive laws hold. Any lattice in which they hold is called ***distributive***.

DEFINITION 2.16.1. A lattice is *modular* if and only if the distributive laws:

$$a \wedge (b \vee c) = (a \wedge b) \vee (a \wedge c)$$

$$a \vee (b \wedge c) = (a \vee b) \wedge (a \vee c)$$

hold whenever some pair of elements of a, b, c is comparable, i.e. some element is less than or equal to another. It is *distributive* if and only if these laws hold for all a, b, c.

We will represent lattices by Hasse diagrams of partially ordered sets, that is a line is drawn from x to y if $x < y$ and no z satisfies $x < z < y$. The higher of x, y in the diagram is the greater.

EXAMPLE 2.16.1.

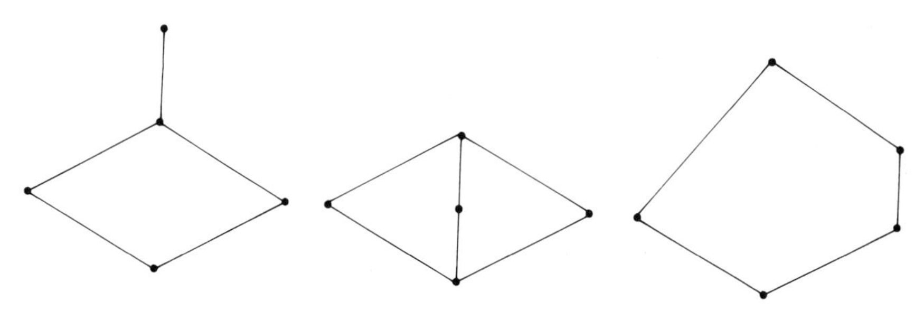

The first lattice is distributive. The second is modular, but not distributive. The third is not modular. The matrices of the three partial orders are

$$\begin{bmatrix} 1 & 0 & 0 & 0 & 0 \\ 1 & 1 & 0 & 0 & 0 \\ 1 & 1 & 1 & 0 & 0 \\ 1 & 1 & 0 & 1 & 0 \\ 1 & 1 & 1 & 1 & 1 \end{bmatrix}, \begin{bmatrix} 1 & 0 & 0 & 0 & 0 \\ 1 & 1 & 0 & 0 & 0 \\ 1 & 0 & 1 & 0 & 0 \\ 1 & 0 & 0 & 1 & 0 \\ 1 & 1 & 1 & 1 & 1 \end{bmatrix}, \begin{bmatrix} 1 & 0 & 0 & 0 & 0 \\ 1 & 1 & 0 & 0 & 0 \\ 1 & 0 & 1 & 0 & 0 \\ 1 & 0 & 1 & 1 & 0 \\ 1 & 1 & 1 & 1 & 1 \end{bmatrix}$$

The latter two lattices are known as a diamond and a pentagon.

A lattice will in fact be modular if and only if $a \wedge (b \vee c) = (a \wedge b) \vee c$ whenever $a \geqslant c$. This law follows from the distributive property $a \wedge (b \vee c) = (a \wedge b) \vee (a \wedge c)$ and is equivalent to it in the only case when it can fail and two elements be comparable, namely $a \leqslant c$. Similarly for the other distributive law.

DEFINITION 2.16.2. A *sublattice* of a lattice L is a subset $M \subset L$ such that M is closed under the operations $\wedge$ and $\vee$.

EXAMPLE 2.16.2. Most subspaces of V_n are not sublattices, because they are not closed under products of vectors. However, the subspace $\{(0, 0), (0, 1), (1, 1)\}$ is a sublattice.

Every sublattice of a modular lattice is modular and every sublattice of a distributive lattice is distributive.

THEOREM 2.16.1. *Any lattice which is not modular contains a sublattice isomorphic to a pentagon.*

Proof. Suppose $a \geqslant c$ but $a \wedge (b \vee c) \neq (a \wedge b) \vee c$. We have $a \geqslant a \wedge b$, $a \geqslant c$ so $a \geqslant (a \wedge b) \vee c$, and $b \vee c \geqslant a \wedge b$, $b \vee c \geqslant c$ so $b \vee c \geqslant (a \wedge b) \vee c$. Thus $a \wedge (b \vee c) \geqslant (a \wedge b) \vee c$. The pentagon will be as follows

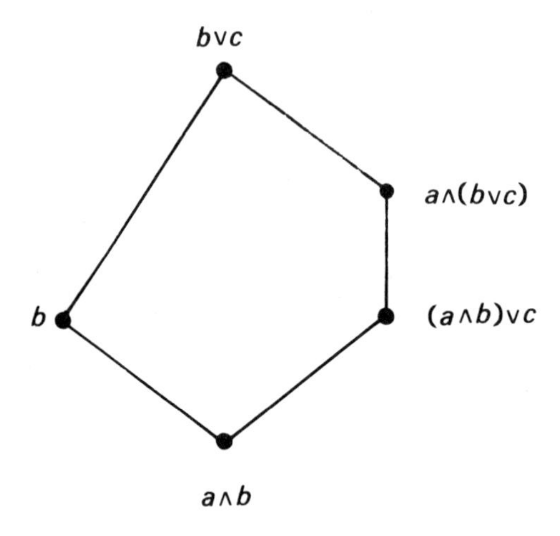

It is immediate that $b \vee c \geqslant b$, $b \vee c \geqslant a \wedge (b \vee c)$, $b \geqslant a \wedge b$, $(a \wedge b) \vee c \geqslant a \wedge b$. We will show that if any of these were equalities, then $a \wedge (b \vee c) \leqslant (a \wedge b) \vee c$. If $b \vee c = b$ then $a \wedge (b \vee c) = a \wedge b$ so this holds. If $b \vee c = a \wedge (b \vee c)$ then $a \geqslant b \vee c$ so $a \wedge b = b$. Thus $a \wedge (b \vee c) = b \vee c = (a \wedge b) \vee c$. If $b = a \wedge b$ then $a \geqslant b$ and so $a \geqslant b \vee c$ again. If $(a \wedge b) \vee c = a \wedge b$ then $c \leqslant a \wedge b$. So $a \wedge (b \vee c) = a \wedge b = (a \wedge b) \vee c$. This shows all the inequalities of the pentagon are strict.

In order that the diagram be a pentagon as partially ordered set, we must not have $b \geqslant (a \wedge b) \vee c$ or $b \leqslant a \wedge (b \vee c)$. In both these cases $b \geqslant c$, or $a \geqslant b$ and we have a contradiction as before. So we have a subset which is a pentagon as partially ordered set.

Finally it remains to show that this set is a sublattice, that it is closed under $\wedge$, $\vee$. We need only consider pairs of elements which are not related by $\leqslant$. For instance $b \vee ((a \wedge b) \vee c) \geqslant b \vee c$. But from the diagram $b \vee ((a \wedge b) \vee c) \leqslant b \vee c$ is immediate. Likewise $b \wedge ((a \wedge b) \vee c) \leqslant b \wedge a$. The reverse inequality follows from the diagram. The other relations follow from these and the diagram. This completes the proof. □

A number of other characterizations of modular lattices are also known (Donellan, 1968).

THEOREM 2.16.2. *Let the normal subgroups of a group G be partially ordered by inclusion. This partial order is a lattice, where $A \vee B = AB$ and $A \wedge B = A \cap B$. Moreover the lattice is modular.*

Proof. Let A, B be normal subgroups. Suppose $a \in A$, $b \in B$. Then $bab^{-1} \in A$. So $bab^{-1} = a_1$, $ba = a_1 b$. Thus $AB \subset BA$. Likewise $BA \subset AB$. Also $a^{-1} \in A$, $b^{-1} \in B$ so $b^{-1}a^{-1} \in AB$. So $ABAB = AABB = AB$ and $(AB)^{-1} \subset AB$. This proves AB is a subgroup of G.

Let $a \in A$, $b \in B$, $g \in G$. Then $gabg^{-1} = gag^{-1}gbg^{-1} \in AB$. So AB is a normal subgroup. Suppose a subgroup $N \supset A$, $N \supset B$. Then $N = NN \supset AB$. Also $AB \supset A$, $AB \supset B$. Thus AB is a least upper bound of A, B.

The set $A \cap B$ is closed under products, inverses, and conjugations, and so is also a normal subgroup. If K is a subgroup, $K \subset A$, $K \subset B$ then $K \subset A \cap B$. Thus $A \cap B$ is a greatest lower bound of A, B.

Thus the normal subgroups form a lattice. Suppose A, B, C are normal subgroups and $A \supset C$. To prove modularity we must show $A \cap (BC) \subset (A \cap B)C$. Let $a \in A$, $a = bc$ where $b \in B$, $c \in C$. Then $b = ac^{-1} \in AC = A$. And $b \in B$. So $B \in A \cap B$ and $c \in C$. Thus $a = bc \in (A \cap B)C$. This proves the theorem. □

There are many unsolved problems in the theory of modular lattices.

In contrast, finite distributive lattices are fairly well known. Every finite lattice has a unique basis under the operation $\vee$, i.e. a minimum generating set (with zero removed). And the structure of this basis as a partially ordered set completely determines the lattice.

EXERCISES

Level 1

1. Why is a linearly ordered set a lattice?
2. Prove a linearly ordered set is distributive (consider six cases for the distributive laws: $a \leqslant b \leqslant c$, $a \leqslant c \leqslant b$, $b \leqslant c \leqslant a$, $c \leqslant b \leqslant a$, $b \leqslant a \leqslant c$, $c \leqslant a \leqslant b$).
3. Draw the lattice of $\{0, 1\}$.
4. Do the same for subsets of $\{0, 1, 2\}$.
5. Draw the lattice of subgroups of $\mathbf{Z}_3 \times \mathbf{Z}_3$. Write out the elements. All subgroups except e, $\mathbf{Z}_3 \times \mathbf{Z}_3$ are cyclic of order 3.
6. Draw the lattice of subgroups of $\mathbf{Z}_8$. (They are all cyclic.)

7. Draw the lattice of subgroups of $\mathbf{Z}_6$. Use the multiplication table. (All subgroups are cyclic.)
8. Show a finite lattice has a smallest element 0, and that $0 \vee x = x$ for all x.

Level 2

1. Find an example of an infinite lattice with no smallest element.
2. A basis is a minimum generating set under $\vee$ of the nonzero elements of a lattice, i.e. a set of generators such that no proper subset generates the lattice. Prove any finite lattice has a basis. Show the set of integers has no basis.
3. Show $\vee$ is associative in any lattice.
4. The positive integers form a lattice under the relation n divides m, although this cannot be proved without some number theory. What is $6 \vee 10$? What is $10 \wedge 6$? Do you think this lattice is distributive?
5. Show that the family of all subgroups of a group (not necessarily normal) is a lattice under inclusion. What are the operations?
6. Show that if L is any finite partially ordered set in which $\wedge$ always exists, and L has an element 1 larger than any other, L is a lattice. Define $\vee$ by $a \vee b = \bigwedge_{z \geqslant a, b} z$.
7. Prove the dual of the above exercise that any subset of a finite lattice containing 0 and closed under $\vee$ is a lattice under the partial order. Give an example for $L =$ subsets of $\{1, 2, 3\}$ where it is not a sublattice.

Level 3

1. Show that if a lattice is modular but not distributive, it contains a diamond.
2. Show that the basis for any finite lattice is unique and consists of all x such that if $x = y + z$ then $x = y$ or $x = z$.
3. Let B be the basis for a finite distributive lattice L regarded as a partially ordered set. Show that B determines the structure of L. Specifically show that L is isomorphic to the lattice of subsets S in B such that if $x \in S, y \leqslant x$ then $y \in S$, that a specific isomorphism is given by sending S to $\bigvee_{x \in S} x$.
4. Prove any distributive lattice with basis B is isomorphic to a sublattice of the lattice of subsets of a set. Show in particular L is isomorphic to a sublattice of the subsets of itself by the 1-to-1 map sending x to $\{y \in B : y \leqslant x\}$. No similar result is known for modular lattices.
5. Prove that if $a \wedge (b \vee c) = (a \wedge b) \vee (a \wedge c)$ for all a, b, c in a lattice, that $a \vee (b \wedge c) = (a \vee b) \wedge (a \vee c)$.
6. For a modular lattice, prove $(x \vee y) \wedge (y \vee z) \wedge (z \vee x) = (x \wedge (y \vee z)) \vee (y \wedge (z \vee x))$. (This identity can be found in Birkhoff (1967).)

CHAPTER 3

Vector spaces

In this chapter we review a number of the most important facts about matrices and linear algebra, some of which are needed in later chapters. A ***field*** is a system such as the rational, real, or complex numbers, in which addition, multiplication, subtraction, and division exist and satisfy the usual laws of algebra (exclusive of order properties). A ***vector space*** over $\mathbf{F}$ is a set $\mathbf{V}$ in which addition $v + w$ is defined and scalar multiplication cv is defined for c in $\mathbf{F}$, v in $\mathbf{V}$, having various properties such as $v + w = w + v$. The standard example is the spaces of n-tuples $\mathbf{F}^n$ of elements of $\mathbf{F}$, where $(x_1, x_2, \ldots, x_n) + (y_1, y_2, \ldots, y_n) = (x_1 + y_1, x_2 + y_2, \ldots, x_n + y_n)$ and $c(x_1, x_2, \ldots, x_n) = (cx_1, cx_2, \ldots, cx_n)$. Vector spaces are used to study systems of linear equations, geometry of n-dimensional space, forces and velocities in physics, supply and demand for many goods in economics, and have other uses.

Every vector space has a set $\{v_1, v_2, \ldots, v_k\}$ of vectors such that all vectors w can uniquely be written as linear combinations $c_1 v_1 + c_2 v_2 + \ldots + c_k v_k$. A set $\{v_1, v_2, \ldots, v_k\}$ is called a ***basis***. The number of vectors $v_1, v_2, \ldots, v_k$ is called the ***dimension*** of the vector spaces.

A ***matrix*** is a rectangular array of numbers. Matrices can be added, subtracted, and multiplied. The rule $AB = BA$ does not always hold, and inverses A^{-1} may not exist. For given bases, there is a 1-1 correspondence between matrices of a certain size and linear transformations f from one vector space to another, i.e. functions f such that $f(cv) = cf(v)$ and $f(v + w) = f(v) + f(w)$. A set of simultaneous linear equations may be written as a single matrix equation $xA = b$.

The ***determinant*** of a matrix is a somewhat complicated function having many properties such as $\det(AB) = \det(A)\det(B)$ and $\det(A) \neq 0$ if and only if A has an inverse. The ***characteristic polynomial*** of an $n \times n$ matrix is the polynomial of degree n $\det(tI - A) = 0$. It is a similarity invariant of a matrix, that is, it is the same for A as for XAX^{-1}. This means it gives propertries of a linear transformation independent of the basis. Its roots are eigenvalues of the matrix A, that is, numbers k such that for a nonzero vector v, $vA = kv$.

For normal matrices, those such that $AB = BA$ where B is the transpose of the complex conjugate of A, more is true. The matrix $A = UDU^{-1}$ where D is a diagonal matrix consisting of eigenvalues of A, and U is a unitary matrix, that is, the transposed complex conjugate of U is U^{-1}. Such matrices represent rotations of the coordinate system. For a symmetric matrix all eigenvalues are real numbers.

3.1 VECTOR SPACES

Vectors are encountered in physics, as line segments whose direction is the direction of a force (or position, velocity, acceleration) and whose length gives the intensity of the force. Such vectors can be added to give the sum of two forces by adding their $x, y,$ and z components (parallelogram method). There also exist methods of multiplying vectors called ***inner*** and ***cross products***.

EXAMPLE 3.1.1. A vector directed upwards at a 45° angle with an intensity of $2\sqrt{2}$ can be represented by $(2, 2)$. Here the line segment from $(0, 0)$ to $(2, 2)$ has direction a 45° angle with the x-axis and length $2\sqrt{2}$.

A vector can be regarded as simply an n-tuple of real numbers. Thus the set of vectors is in 1-1 correspondence with the set of points in n-dimensional space, represented by coordinates. Unlike points, however, vectors can be added and multiplied by numbers.

EXAMPLE 3.1.2. $(1, 2, 7) + (-4, 2, 1) = (-3, 4, 8)$.

EXAMPLE 3.1.3. $2(1, 2, 3) = (2, 4, 6)$.

This viewpoint is useful in various problems involving linear equations, such as the solution of simultaneous linear equations. The solutions for all variables $x_1, x_2, \ldots, x_n$ can be regarded as a single vector instead of many distinct quantities. In economics quantities of many different foods are often regarded as a single vector. For instance $(2, 3, 10)$ might denote 2 bushels of wheat, 3 pounds of steel, and 10 grams of gold. Vectors can indicate different aspects of a sensation, such as intensity and pitch of sound.

A more general viewpoint is taken in abstract algebra. Here a vector is anything, not necessarily consisting of numbers on which operations of addition and multiplication by elements of another set as in Example 3.1.2, can be defined.

A ***field*** is a set like the real numbers obeying the usual laws of algebra, in which we can divide by any nonzero element.

DEFINITION 3.1.1. A *field* is a set $\mathbf{F}$ on which binary operations addition and multiplication $\mathbf{F} \times \mathbf{F} \to \mathbf{F}$ are defined, with these properties:

$$(a + b) + c = a + (b + c), \qquad (ab)c = a(bc)$$

$$a + b = b + a, \qquad ab = ba$$

$$a(b + c) = ab + ac$$

There exist 0, 1 such that for all x, $x + 0 = x$, $x \cdot 1 = x$.

For any x there exists $-x$ and if $x \neq 0$ there exists $\frac{1}{x}$ such that $-x + x = 0$, $x \cdot \frac{1}{x} = 1$

EXAMPLE 3.1.4. The rational, real, and complex numbers are fields.

EXAMPLE 3.1.5. The set of numbers of the form $a + b\sqrt{2}$, $a, b \in \mathbf{Q}$ is a field.

The field will be the set of 'numbers' by which vectors can be multiplied as $3(1, -1) = (3, -3)$. Elements of it are called *scalars*.

DEFINITION 3.1.2. A *vector space* over a field $\mathbf{F}$ is a set $\mathbf{V}$ with an operation of scalar multiplication $\mathbf{F} \times \mathbf{V} \to \mathbf{V}$ and an operation of addition $\mathbf{V} \times \mathbf{V} \to \mathbf{V}$ such that the following are true for all $s, t \in \mathbf{F}$, $v, w, z \in \mathbf{V}$ and for a fixed $0 \in \mathbf{V}$:

$$s(tv) = (st)v, \qquad 1v = v$$

$$(s + t)v = sv + tv, \qquad s(v + w) = sv + sw$$

$$(v + w) + z = v + (w + z), \qquad v + w = w + v$$

$$0 + v = v$$

and, for the zero, identity elements 0, 1 of the field we have

$$0 \cdot v = 0, \qquad 1 \cdot v = v$$

It follows that $(-1)v + v = (-1)v + 1 \cdot v = (-1 + 1)v = 0 \cdot v = 0$ so that a vector space is an abelian group under addition.

EXAMPLE 3.1.6. The $\mathbf{F}$ is itself a vector space over $\mathbf{F}$.

EXAMPLE 3.1.7. The set of all n-tuples $(v_1, v_2, \ldots, v_n)$ over $\mathbf{F}$ is a field with operations $(x_1, x_2, \ldots, x_n) + (y_1, y_2, \ldots, y_n) = (x_1 + y_1, x_2 + y_2, \ldots, x_n + y_n)$ and $c(v_1, v_2, \ldots, v_n) = (cv_1, cv_2, \ldots, cv_n)$, $c \in \mathbf{F}$.

EXAMPLE 3.1.8. The $M_n(F)$ is a vector space.

EXAMPLE 3.1.9. The set of functions from any set X to $\mathbf{F}$ is a vector space under the usual addition of functions and multiplication of functions by constants.

DEFINITION 3.1.3. A *subspace* of a vector space $\mathbf{V}$ is a subset of $\mathbf{V}$ closed under addition and under multiplication by field elements.

EXAMPLE 3.1.10. The set of polynomial functions from $\mathbf{R}$ to itself is a subspace of the vector space of all functions from $\mathbf{R}$ to itself.

The concept of linear dependence may be recalled from the theory of simultaneous linear equations. A set of linear equations is dependent if one is a linear combination of the others. In the same way a set of vectors is linearly dependent if one is a linear combination of the others.

DEFINITION 3.1.4. Let $\mathbf{I}$ be a set, called an *index set*. Let f be a function from $\mathbf{I}$ to a family of sets. For $i \in \mathbf{I}$, let the values of f be denoted A_i. Then the sets A_i form what is called an *indexed family of sets*. The product of the indexed family of sets A_i is the set of all functions g from $\mathbf{I}$ to the union A of all sets A_i, such that $g(i) \in A_i$ for all i.

EXAMPLE 3.1.11. Let $\mathbf{I} = \{1, 2, 3\}$. Then our indexed family consists of three sets A_1, A_2, A_3. The product is the set of all functions g such that $g(1) \in A_1$, $g(2) \in A_2$, $g(3) \in A_3$. This can be written $\{(g_1, g_2, g_3) : g_1 \in A_1, g_2 \in A_2, g_3 \in A_3\}$. So we see this really is the same as the previously defined Cartesian product.

DEFINITION 3.1.5. A set S in a vector space $\mathbf{V}$ is *linearly dependent* if there exist elements $f_1, f_2, \ldots, f_n \in \mathbf{F}$, not all zero, and distinct elements $v_1, v_2, \ldots, v_n \in \mathbf{V}$ for some positive integer n such that $f_1 v_1 + f_2 v_2 + \ldots + f_n v_n = 0$. Otherwise it is *linearly independent*. An indexed sequence $\{v_i\}_{i \in \mathbf{I}}$ of vectors is linearly independent if all the vectors v_i are distinct and the set $\{v_i\}_{i \in \mathbf{I}}$ is independent. Here 0 denotes the zero vector.

EXAMPLE 3.1.12. The set of vectors $\{(1, 0, 0), (1, 0, -1), (-1, 0, 2)\}$ is linearly dependent, since

$$(-1)(1, 0, 0) + 2(1, 0, -1) + (1)(-1, 0, 2) = (0, 0, 0)$$

EXAMPLE 3.1.13. Any set containing the vector 0 is dependent.

EXAMPLE 3.1.14. The set $\{v, 2v\}$ is dependent.

EXAMPLE 3.1.15. The set $\{v, w, v + w\}$ is dependent.

EXAMPLE 3.1.16. For any n the set of n-tuples $\{(1, 0, \ldots, 0), (0, 1, \ldots, 0), \ldots, (0, 0, \ldots, 1)\}$ with $(n - 1)$ zeros and 1 one is linearly independent.

DEFINITION 3.1.6. The *span* $\langle S \rangle$ of a set S of vectors is the set of all linear combinations $f_1 v_1 + f_2 v_2 + \ldots + f_n v_n$ where $f_i \in \mathbf{F}$, $v_i \in \mathbf{V}$, and n is any positive integer.

EXAMPLE 3.1.17. The span of $\{(1, 0, 0), (0, 1, 0)\}$ is the set of all vectors of the form $(f_1, f_2, 0)$ for $f_i \in \mathbf{F}$.

If a set of vectors is linearly dependent, some vector v will be a linear combination of the others. Take any v_i such that $f_i \neq 0$ in the equation $f_1 v_1 + f_2 v_2 + \ldots + f_n v_n = 0$ and solve for v_i in terms of the rest. If a vector is a linear combination of others, removing it will not affect the span, since we can substitute this expression for it in any other linear combination.

PROPOSITION 3.1.1. *The span of S is a subspace and is contained in every subspace containing S.*

Proof. A sum of linear combinations is again a linear combination, and a multiple of a linear combination by an element of $\mathbf{F}$ is again a linear combination. Therefore $\langle S \rangle$ has the closure properties needed to be a subspace. If $S \subset W$ for a subspace W then by closure W must contain all linear combinations of elements of S, i.e. it contains $\langle S \rangle$. □

Vector spaces are frequently studied in terms of subsets called *bases*. Two vector spaces with bases of the same size are isomorphic. A basis gives a system of coordinates (which need not be at right angles to one another). Thus the usual x, y, z coordinates correspond to the basis $\{(1, 0, 0), (0, 1, 0), (0, 0, 1)\}$.

DEFINITION 3.1.7. A *basis* for $\mathbf{V}$ is an independent subset $S \subset \mathbf{V}$ that spans $\mathbf{V}$.

PROPOSITION 3.1.2. *The set $\{v_1, v_2, \ldots, v_n\}$ is a basis for* $\mathbf{V}$ *if and only if every vector $v \in \mathbf{V}$ has a unique expression $v = f_1 v_1 + f_2 v_2 + \ldots + f_n v_n$ where $f_i \in \mathbf{F}$.*

Proof. Let $\{v_1, v_2, \ldots, v_n\}$ be a basis. Since it is a spanning set, for any v, $v = f_1 v_1 + f_2 v_2 + \ldots + f_n v_n$ for some $f_i \in \mathbf{F}$. If the f_i are not unique take the difference of two expressions for v. This gives a relation $g_1 v_1 + g_2 v_2 + \ldots + g_n v_n = 0$ where not all $g_i = 0$, a contradiction to independence.

Suppose conversely every vector $v \in \mathbf{V}$ has a unique expression $v = f_1 v_1 + f_2 v_2 + \ldots + f_n v_n$. Then $v_1, v_2, \ldots, v_n$ span $\mathbf{V}$. Let $v = 0$. Uniqueness shows that all $f_i = 0$. This proves independence. □

EXAMPLE 3.1.18. The set $\{(1, 0), (0, 1)\}$ is a basis for any fixed $a \in \mathbf{F}$, for the space of all ordered pairs (f_1, f_2).

THEOREM 3.1.3. *Any independent set maximal under inclusion is a basis, and any spanning set minimal under inclusion is a basis. Every independent set is contained in a basis and every spanning set contains a basis.*

Proof. Suppose S spans $\mathbf{V}$ but is not a basis. Then some vector $v \in S$ is a linear combination of the other members of S. Thus if w is deleted, we still have a spanning set. So S is not minimal.

Suppose S is independent but not a basis. Then it does not span S. So some vector v of $\mathbf{V}$ is not a linear combination of elements of S. Then $S \cup \{v\}$ is independent: if $f_1 v_1 + f_2 v_2 + \ldots + f_n v_n + f_{n+1} v = 0$ then $f_{n+1} v = 0$ since $v \notin \langle S \rangle$ and $f_i = 0, i < n+1$ since S is independent. This proves S is not maximal.

Let S be an independent set. Consider the family $\mathcal{F}$ of independent sets which contain S. The family $\mathcal{F}$ is partially ordered by inclusion. Any chain C in $\mathcal{F}$ is bounded by the union of the members of C, which will be independent: if $f_1 v_1 + f_2 v_2 + \ldots + f_n v_n = 0$ take members $C_1, C_2, \ldots, C_n$ of C containing $v_1, v_2, \ldots, v_n$ respectively. Let C_k be the largest under inclusion. Then all $v_i \in C_k$ and $f_i = 0$ by independence of C_k. Thus C is bounded. So by Zorn's Lemma, $\mathcal{F}$ has a maximal element. This will be a basis by what was proved above.

Let S be a spanning set. The Well-Ordering Principle which can be proved from the Axiom of Choice, states that there exists a linear ordering $<$ of S in which every subset of S has a least element.

Let $T = \{s \in S : s$ is not a linear combination of elements $t < s$ in the ordering$\}$. Suppose T is dependent. Let $f_1 t_1 + f_2 t_2 + \ldots + f_n t_n = 0$. Assume all $f_i \neq 0$ else delete that term. Let the last of $t_1, t_2, \ldots, t_n$ be t_k. Then t_k is a linear combination of prior elements. This contradicts the definition of T.

Suppose T does not span $\mathbf{V}$. Then since $\langle S \rangle = \mathbf{V}$ it cannot span S. Let t be the first member of S not in $\langle T \rangle$. Then t is not a linear combination of prior elements since prior elements are in $\langle T \rangle$ and t would be. So by definition of t, $t \in T$. This is a contradiction. So T is a basis. □

COROLLARY 3.1.4. *Every vector space has a basis.*

This theorem assumes the Axiom of Choice, and cannot be proved without it for infinite-dimensional spaces. For finite-dimensional spaces, its main use, the Axiom of Choice is not needed.

EXAMPLE 3.1.19. The independent set $\{(1,0,1),(2,1,0)\}$ is included in the basis $\{(1,0,1),(2,1,0),(0,0,1)\}$.

EXAMPLE 3.1.20. The set $S = \{(0,0,0),(1,0,1),(2,0,2),(0,1,0),(1,1,1),(0,0,1),(1,0,2)\}$ spans the set of all ordered triples of $\mathbf{R}$. To obtain a basis delete all elements of S which are linear combinations of prior elements. This leaves $\{(1,0,1),(0,1,0),(0,0,1)\}$ which is a basis.

We will conclude this section with a brief history of vector analysis. The parallelogram method of addition for vectors was discovered by the Greeks for addition of velocities. It was used to represent addition of forces by physicists in the sixteenth and seventeenth centuries. C. Wessel, a Norwegian surveyor, in 1799 discovered the geometric representation of complex numbers (which are a vector space over **R**) and extended some results to 3 dimensions. This result was independently discovered by C. F. Gauss, J. R. Argand, C. Mourey, and J. Warren. W. R. Hamilton tried for years to find a system that extends the complex numbers in the same way the complex numbers extend the real numbers. He found no 3-dimensional system, but a 4-dimensional noncommutative one called the ***quaternions***, with basis 1, $\boldsymbol{i}, \boldsymbol{j}, \boldsymbol{k}$ where $\boldsymbol{i}^2 = \boldsymbol{j}^2 = \boldsymbol{k}^2 = -1$ and $\boldsymbol{ij} = \boldsymbol{k}$, $\boldsymbol{jk} = \boldsymbol{i}$, $\boldsymbol{ki} = \boldsymbol{j}$, $\boldsymbol{ji} = -\boldsymbol{k}$, $\boldsymbol{kj} = -\boldsymbol{i}$, $\boldsymbol{ik} = -\boldsymbol{j}$. This system can be used to represent rotations in 3-dimensional space. He also introduced the operators del and del^2 of mathematical physics.

Operations of dot and cross product were introduced by A. M. Moebius, H. G. Grassman, A. Barre and others. Grassmann in particular considered very general types of n-dimensional systems with vector space operations and various types of products, some not even associative.

P. G. Tait further developed the theory of quaternions. The modern form of the theory of vectors is due to two mathematical physicists, J. W. Gibbs and O. Heaviside. Much of the above material is from J. M. Crowe (1967).

The nineteenth century was the time of the beginnings of a number of other concepts of abstract algebra. A. Cauchy studied cycles of permutations and transpositions. E. Galois studied groups as a method of proving polynomials of the fifth degree were not solvable, and discovered the importance of normal subgroups and commutativity. A. Cayley proved his previous cited theorem.

Fields of algebraic numbers were implicit in the work of E. Galois on solving polynomials. Finite fields $\mathbf{Z_p}$ were discovered by C. F. Gauss, in his work on congruences. Gauss also discovered what in effect were deep results about algebraic number fields in his work on quadratic forms, quadratic reciprocity, and other types of number theory. E. Kummer studied algebraic number fields in order to solve many cases of Fermat's still unsolved conjecture that $x^n + y^n = z^n$, $n \geqslant 3$ has no solutions in whole numbers. These authors did not, however, use the abstract, axiomatic approach based on set theory which is used today. This came about as an attempt to find rigorous foundations for all of mathematics, especially calculus, in the late nineteenth century. G. Cantor is generally considered the inventor of set theory.

EXERCISES

Level 1

1. Add $(1, 2, 3), (4, -5, 6), (-5, -10, 1)$.
2. Multiply $(1, 0, -2)$ by -3.

3. If $x = (1, 2, 3)$, $y = (1, 0, 1)$, $z = (0, 0, 1)$ write $2x + 3y + 4z$.
4. Find a, b such that $(1, 2) = a(1, 0) + b(1, 1)$.
5. Is this set dependent? $\{(1, 1), (2, 3), (0, 1)\}$. Find an independent subset.

Level 2

1. Prove that for any a, b, c: $\{(1, 0, 0), (a, 1, 0), (b, c, 1)\}$ gives a basis for all triples of real numbers.
2. Give an example of a subset of $\{(a, b): a, b \in \mathbf{R}\}$ which is closed under addition but is not a vector space over $\mathbf{R}$.
3. Tell why if a field $\mathbf{F}_1$ is contained in $\mathbf{F}_2$ that $\mathbf{F}_2$ is a vector space over $\mathbf{F}_1$. What is a basis for the complex numbers as a vector space over $\mathbf{R}$?
4. Find a subset of this set which is a basis, by the method of Example 3.1.20: $\{(0, 1, 0), (1, 2, 0), (2, 4, 0), (1, 3, 0), (1, 1, 1), (2, 2, 2)\}$.
5. Extend this independent set to a basis: $\{(1, 1, 1), (1, 2, 0)\}$. To do this, add a spanning set $\{(1, 0, 0), (0, 1, 0), (0, 0, 1)\}$ and apply the method of the last exercise.

Level 3

1. Does any countable set have a well-ordering?
2. Use the existence of a basis for the real numbers over the rationals to construct a function f from the real numbers to itself such that $f(x + y) = f(x) + f(y)$ but f is not continuous.
3. Add the quaternions $1 + 2\mathbf{i} + 3\mathbf{j} + 4\mathbf{k}$ and $2 + \mathbf{i} + \mathbf{j} - \mathbf{k}$.
4. Multiply them. The distributive law holds for quaternions.
5. Prove that the associative law holds for the quaternions. It suffices to check the basis elements $\pm 1, \pm \mathbf{i}, \pm \mathbf{j}, \pm \mathbf{k}$. If one of x, y, z is the identity, the relation $(xy)z = x(yz)$ always holds. Factor out all minus signs. By symmetry in $\mathbf{i}, \mathbf{j}, \mathbf{k}$ assume $x = \mathbf{i}$. Then check 9 cases for $y, z = \mathbf{i}, \mathbf{j}, \mathbf{k}$.

3.2 BASIS AND DIMENSION

In geometry a point is said to have dimension 0, a line dimension 1, a plane dimension 2, and all of three-dimensional space dimension 3. These can be interpreted as the number of continuous real coordinates needed to give the position of a point, or the number of simultaneously perpendicular line segments which can exist. The measures length, area, volume are related to this.

Dimension generalizes to vector spaces. Instead of coordinates we simply consider the number of vectors in a basis. However, we must show this number is the same for all bases.

First we state a frequently used result about simultaneous linear equations, valid also in the case of an infinite number of equations and variables.

As previously mentioned two sets S, T are said to have the same cardinality if there exists a 1-1 onto mapping from S to T.

DEFINITION 3.2.1. The *cardinality of S is less than or equal to that of T,* writren $|S| \leqslant |T|$ if and only if there exists a 1-1 mapping from S into T.

EXAMPLE 3.2.1. For finite sets this means the number of elements in S is less than or equal to that in T.

EXAMPLE 3.2.2. Since $\mathbf{Z} \subset \mathbf{R}$, $|\mathbf{Z}| \leqslant |\mathbf{R}|$.

Assuming the Axiom of Choice this definition is equivalent to saying there exists an onto mapping from T to S. It is readily verified that the relation $|S| \leqslant |T|$ is reflexive and transitive (compose two 1-1 maps $S \rightarrow T$ and $T \rightarrow U$). The result that if $|S| \leqslant |T|$ and $|T| \leqslant |S|$ then $|T| = |S|$ is less obvious and is called the ***Schröder–Bernstein Theorem***. In addition we use the fact that for S infinite, $|S| = |S| \times |\mathbf{Z}|$.

Linear equations are called ***homogeneous*** if they have no constant terms except zero. For instance $2x + 3y = 0$ is homogeneous but not $2x + 3y = 1$. Coefficients are assumed to be in $\mathbf{F}$.

THEOREM 3.2.1. *Consider a set E of homogeneous linear equations in distinct variables $\{x_i\}_{i \in \mathbf{I}}$. If $|\mathbf{I}| > |E|$ then there exists a solution of all equations in which the set of nonzero variables is finite and nonempty.*

Proof. Each equation can contain only a finite number of variables. Thus if E is infinite the number of variables which actually appear in the equations will be no more than $|\mathbf{Z} \times E| = |E| < |\mathbf{I}|$. So some variables will not appear in any of the equations. Let the variables which appear in the equations be zero, and assign arbitrary nonzero values to finitely many others.

Thus we may assume E is finite. We may now by a standard procedure reduce E to a simpler system E'. Take the first equation and choose a variable which occurs in it with nonzero coefficient. Eliminate this variable from the other equations by adding or subtracting a multiple of the first equation. At the ith stage, if the ith equation is zero, delete it. Otherwise choose a variable which occurs in it with nonzero coefficient, and eliminate that variable from all the other equations by adding or subtracting a multiple of the ith equation. At the final stage each chosen variable will occur in only one equation. Since there are more variables than equations, some variables must not have been chosen. Assign variables not chosen, arbitrary nonzero values. Then there will exist unique values of the chosen variables satisfying all equations (just solve the equations for the chosen variables). □

EXAMPLE 3.2.3. The system

$$\begin{cases} x + 2y + z = 0 \\ 3x - 2y + 4z = 0 \end{cases}$$

has the nonzero solution $x = -10$, $y = 1$, $z = 8$. Any multiple of this gives another solution.

THEOREM 3.2.2. *Any two bases of a vector space have the same cardinality.*

Proof. Let $\{v_i\}_{i \in \mathbf{I}}$ and $\{w_j\}_{j \in \mathbf{J}}$ be two bases for $\mathbf{V}$. Suppose $|\mathbf{I}| < |\mathbf{J}|$. Since $\{v_i\}_{i \in \mathbf{I}}$ is a spanning set, there exist $a_{ij} \in \mathbf{F}$

$$w_j = \sum_{i \in \mathbf{I}} a_{ij} v_i$$

for each $j \in \mathbf{J}$. By Theorem 3.2.1 there exist $\{x_j\}_{j \in \mathbf{J}}$ not all zero but only finitely many nonzero solving the equations

$$\sum_{j \in \mathbf{J}} a_{ij} x_j = 0$$

for all $i \in \mathbf{I}$. Therefore

$$\sum_{j \in \mathbf{J}} x_j w_j = \sum_{j \in \mathbf{J}} \sum_{i \in \mathbf{I}} x_j a_{ij} v_i = \sum_{i \in \mathbf{I}} v_i \sum_{j \in \mathbf{J}} x_j a_{ij}$$

$$= \sum_{i \in \mathbf{I}} v_i(0) = 0.$$

Thus the w_j are linearly dependent. This is a contradiction. By symmetry, $|\mathbf{I}| > |\mathbf{J}|$ is also impossible. So $|\mathbf{I}| = |\mathbf{J}|$. □

The Cartesian product of a collection of vector spaces is itself a vector space in a natural way.

DEFINITION 3.2.2. The (*external*) *direct sum* of vector spaces $\mathbf{V_1}, \mathbf{V_2}, \ldots, \mathbf{V_n}$ over $\mathbf{F}$ is the set of all n-tuples $(v_1, v_2, \ldots, v_n)$ such that $v_i \in \mathbf{V_i}$ for each i. Addition and scalar multiplication are defined by $(v_1, v_2, \ldots, v_n) + (w_1, w_2, \ldots, w_n) = (v_1 + w_1, v_2 + w_2, \ldots, v_n + w_n)$, $c(v_1, v_2, \ldots, v_n) = (cv_1, cv_2, \ldots, cv_n)$ where $c \in \mathbf{F}$.

EXAMPLE 3.2.4. The direct sum $\mathbf{F} \oplus \mathbf{F} \oplus \ldots \oplus \mathbf{F}$ is the usual space of n-tuples. It is denoted $\mathbf{F^n}$. This vector space is the same as the Cartesian product $\mathbf{F^n}$ mentioned in previous sections.

This definition extends to infinite direct sums and products. The ***direct product*** is as in Definition 3.1.4. A direct sum, or restricted direct product in somewhat confusing language is the subset of the direct product consisting of functions which have only a finite number of nonzero values.

DEFINITION 3.2.3. The *dimension* of a vector space is the cardinality of a basis.

EXAMPLE 3.2.5. The dimension of $\mathbf{C}$ over the field $\mathbf{R}$ is 2, since $\{1, i\}$ is a basis.

A basis for a direct sum space can be obtained as the union of bases for each vector space $\mathbf{V_i}$. Therefore the dimension of a direct sum of vector spaces $\mathbf{V_i}$ is the sum of dim $\mathbf{V_i}$.

EXAMPLE 3.2.6. The dimension of $\mathbf{F^n}$ is n.

The expression 'internal direct sum' means that a vector space is isomorphic to a direct sum of certain of its own subspaces, rather than 'outside' vector spaces.

DEFINITION 3.2.4. A vector space $\mathbf{V}$ is the *internal direct sum* of subspaces $W_1, W_2, \ldots, W_n$ if and only if every vector $v \in \mathbf{V}$ can be uniquely expressed as $v = w_1 + w_2 + \ldots + w_n$, with $w_i \in W_i$. It is denoted $W_1 \oplus W_2 \oplus \ldots \oplus W_n$.

EXAMPLE 3.2.7. The vector space $\mathbf{F}^2$ is the direct sum of the subspaces $W_1 = \{(x, x) : x \in \mathbf{F}\}$ and $W_2 = \{(x, 0) : x \in \mathbf{F}\}$. In fact $(x, y) = (y, y) + (x - y, 0)$ for all $x, y \in \mathbf{F}$.

PROPOSITION 3.2.3. *An internal direct sum is isomorphic to the external direct sum of the same subspaces.*

Proof. Let $\mathbf{V}$ be the internal direct sum of $W_1, W_2, \ldots, W_n$. Let $f : \mathbf{V} \to W_1 \oplus W_2 \oplus \ldots \oplus W_n$ be defined by $f(w_1 + w_2 + \ldots + w_n) = (w_1, w_2, \ldots, w_n)$. Let $g : W_1 \oplus W_2 \oplus \ldots \oplus W_n \to \mathbf{V}$ be defined by $g(w_1, w_2, \ldots, w_n) = w_1 + w_2 + \ldots + w_n$. Then f and g are homomorphisms, and $gf(x) = x$ and $fg(y) = y$. Thus g is the inverse of f. □

DEFINITION 3.2.5. If $W_1, W_2, \ldots, W_n$ are subspaces of $\mathbf{V}$, the *sum* $W_1 + W_2 + \ldots + W_n$ denotes $\{w_1 + w_2 + \ldots + w_n : w_i \in W_i \text{ for each } i\}$.

EXAMPLE 3.2.8. If $W_1 = \{(x, y, 0) : x, y \in \mathbf{F}\}$ and $W_2 = \{(0, y, z) : y, z \in \mathbf{F}\}$ then $W_1 + W_2$ is the set of all triples $\{(x, y, z) : x, y, z \in \mathbf{F}\}$.

PROPOSITION 3.2.4. *Let* $W_1, W_2, \ldots, W_n$ *be the subspaces of* $\mathbf{V}$. *Then* $W_1 + W_2 + \ldots + W_n = \langle W_1 \cup W_2 \cup \ldots \cup W_n \rangle$.

Proof. Every element of $w_1 + w_2 + \ldots + w_n$ is a linear combination of elements of w_i. Conversely every linear combination of elements $w_i \in W_i$ is a sum of elements $t_i w_i \in W_i$, $t_i \in \mathbf{F}$. □

DEFINITION 3.2.6. The subspace W_1 of $\mathbf{V}$ is a *complement* of the subspace W_2 if and only if $W_1 \cap W_2 = \{0\}$ and $W_1 + W_2 = \mathbf{V}$.

EXAMPLE 3.2.9. The subspace $\{(x, 0, 0) : x \in \mathbf{F}\}$ is a complement of $\{(0, y, z) : y, z \in \mathbf{F}\}$.

PROPOSITION 3.2.5. *Let W_1, W_2 be the subspaces of $\mathbf{V}$. Then the subspace W_1 is a complement of the subspace W_2 if and only if $\mathbf{V} = W_1 \oplus W_2$.*

Proof. If $\mathbf{V} = W_1 \oplus W_2$ and $W_1 \cap W_2 \neq \emptyset$ then $x \in W_1 \cap W_2$ can be expressed in two ways: $x + 0$ and $0 + x$. Conversely if $u_1 + u_2 = w_1 + w_2$ where $u_i, w_i \in W_i$ then $u_1 - w_1 = w_2 - u_2 \in W_1 \cap W_2$. So $W_1 \cap W_2 \neq \{0\}$. □

THEOREM 3.2.6. *Any subspace of a vector space has a complement.*

Proof. Let $W \subset \mathbf{V}$. Let B be a basis for W. Then B is an independent set in $\mathbf{V}$. Thus by Theorem 3.1.3 there exists a basis C of $\mathbf{V}$ such that $B \subset C$. Let $D = C \backslash B$. Then $\langle B \rangle + \langle D \rangle = \mathbf{V}$. Let $v \in \langle B \rangle \cap \langle D \rangle$. Then $v = \Sigma\, a_i u_i$, $u_i \in B$ and $v = \Sigma\, b_i w_i$, $w_i \in D$. So $0 = \Sigma\, a_i u_i + (-\Sigma\, b_i w_i)$. Since $B \cup D = C$ is linearly independent, all $a_i = 0$ and all $b_i = 0$. So $v = 0$. Thus $\langle D \rangle$ is a complement of $\langle B \rangle = W$. □

For a vector space $\mathbf{V}$ and subspace W the relation $\{(x, y) : x - y \in W\}$ is a congruence, that is, an equivalence relation such that if $x \sim y$ then $ax \sim ay$ and $x + z \sim y + z$. Therefore the set of equivalence classes forms a vector space under the same operations. This vector space of equivalence classes is called the ***quotient vector space***, denoted $\mathbf{V}/W$. For any complement U of W, the mapping $U \to \mathbf{V} \to \mathbf{V}/W$ is an isomorphism. It is onto since $u \sim u + w$ if $w \in W$ and 1–1 since the kernel is $W \cap U = \{0\}$. Since $\mathbf{V} = W \oplus U$, dim $\mathbf{V}$ = dim W + dim U. Therefore dim $\mathbf{V}/W$ = dim U = dim $\mathbf{V}$ − dim W.

Thm or Cor.

EXERCISES

Level 1

1. Find a nonzero solution of these equations:

$$\begin{cases} 2x + 3y + z = 0 \\ x + 4y - 2z = 0 \end{cases}$$

2. Find an example of 2 nonhomogeneous linear equations with no solution, in 3 variables.
3. What is the dimension of the space of vectors $\{(x, y, -x, -y)\}$? Show $\{(1, 0, -1), (0, 1, -1)\}$ is a basis.
4. If $\mathbf{F}$ is a finite field how many elements does a vector space of dimension n have?
5. Find a complement to the space spanned by $\{(1, 1)\}$ in $\mathbf{R}^2$. (Extend $\{(1, 1)\}$ to a basis.)

Level 2

1. Prove $|\mathbf{Z} \times \mathbf{Z}| \leqslant |\mathbf{Z}|$ because $|\mathbf{Z}| = |\mathbf{Z}^+|$ and the map $(x, y) \to 2^x 3^y$ on $\mathbf{Z}^+$ is 1-1. Here $\mathbf{Z}^+$ denotes the set of positive integers.
2. Prove $|\mathbf{R} \times \mathbf{R}| \leqslant |\mathbf{R}|$ as follows: write a real number x as an infinite decimal (if $a999\ldots$ occurs replace it by $a + 10\,000$). For two real numbers x, y. form the real numbers obtained by alternating digits of x, y. Thus if $x = 0.101$ and $y = 0.345$ send (x, y) to 0.130415. Show $|\mathbf{R} \times \mathbf{Z}| \leqslant |\mathbf{R}|$.
3. What is the dimension of $\{(x, y, z) : x + 2y + 3z = 0\}$? Find a basis.
4. What is the dimension of $\{(x, y, z) : x + y = 0, y + z = 0\}$? Find a basis.
5. Find a complement to the space spanned by $(1, 1, -2)$ in $\{(x, y, z) : x + y + z = 0\}$.

Level 3

1. Prove the Schröder-Bernstein Theorem in the case where one set is finite.
2. Using Exercise 1 prove when both sets have cardinality at most $|\mathbf{Z}|$, hence have cardinality exactly $|\mathbf{Z}|$ or are finite.
3. Find (possibly in a book) a general proof of the Schröder-Bernstein Theorem. One method is as follows: Suppose we have 1-1 maps $f: X \to Y$ and $g: Y \to X$. To an element $x \in X$ associate the set C_x of all elements of X which can be obtained from x by iterating f, g, f^{-1}, g^{-1} where defined. For $y \in Y$ define a similar set D_y. Show that $f(C_x) \subset D_{f(x)}$ and $g(D_y) \subset C_{g(x)}$. This gives a 1-1 correspondence between the family $\{C_x\}$ and the family $\{D_x\}$. To construct a 1-1 correspondence from X to Y it suffices to construct a 1-1 correspondence on each C_x to $D_{f(x)}$ by the Axiom of Choice. This reduces the problem from X, Y to the case $C_x, D_{f(x)}$. Show these sets are at most countable. Next use Exercise 2.
4. For vector spaces $\mathbf{U}, \mathbf{V}, \mathbf{W}$, let i_1, i_2 be the inclusion $\mathbf{U} \to \mathbf{U} \oplus \mathbf{V}, \mathbf{V} \to \mathbf{U} \oplus \mathbf{V}$ and π_1, π_2 the maps $\pi_1(x, y) = x$, $\pi_2(x, y) = y$. For any vector space homomorphisms f_1, f_2 (i.e. $f(ax) = af(x)$ and $f(x + y) = f(x) + f(y)$) from $\mathbf{W} \to \mathbf{U}$, $\mathbf{W} \to \mathbf{V}$, show there exists a unique homomorphism $f(\mathbf{W}) \to \mathbf{U} \oplus \mathbf{V}$ such that $f \circ \pi_1 = f_1$, $f \circ \pi_2 = f_2$. Show for any homomorphisms $g_1: \mathbf{U} \to \mathbf{W}$ and $g_2 : \mathbf{V} \to \mathbf{W}$ there exist a unique homomorphism $g : \mathbf{U} \oplus \mathbf{V} \to \mathbf{W}$ such that $i_1 \circ g = g_1$, $i_2 \circ g_2 = g_2$.
5. Prove either property in Exercise 4 uniquely characterizes $\mathbf{U} \oplus \mathbf{V}$, that for any space $\mathbf{W}$ having the property $\mathbf{W}$ must be isomorphic to $\mathbf{U} \oplus \mathbf{V}$.

3.3 MATRICES

It is assumed that the reader is familiar with basic operations on matrices by earlier study but we review the basic definitions here.

An $n \times m$ matrix over a set S is an indexed collection a_{ij} of elements of S for $i = 1, 2, \ldots, n$ and $j = 1, 2, \ldots, m$. That is, it is a function from the set $\{1, 2, \ldots, n\} \times \{1, 2, \ldots, m\}$ to S. Matrices are usually represented as rectangles of elements of S, with the value a_{ij} in row i and column j. However, they exist as functions independent of any particular diagram. The set S will usually be assumed to be a field, or at least to have operations of addition and multiplication in it, as well as additive and multiplicative identities 0, 1.

DEFINITION 3.3.1. The $n \times n$ *identity matrix* is the matrix $I = (\delta_{ij})$ such that $\delta_{ij} = 1$ if $i = j$ and $\delta_{ij} = 0$ if $i \neq j$.

EXAMPLE 3.3.1.

$$\begin{bmatrix} 1 & 0 & 0 \\ 0 & 1 & 0 \\ 0 & 0 & 1 \end{bmatrix}$$

DEFINITION 3.3.2. The $m \times n$ *zero matrix* 0 is the matrix all of whose entries are 0.

EXAMPLE 3.3.2.

$$\begin{bmatrix} 0 & 0 & 0 & 0 \\ 0 & 0 & 0 & 0 \\ 0 & 0 & 0 & 0 \end{bmatrix}$$

DEFINITION 3.3.3. *Addition* and *multiplication* of matrices are defined by $(A + B)_{ij} = (a_{ij} + b_{ij})$ and $(AB)_{ij} = (\sum_k a_{ik} b_{kj})$.

This means in the case of addition that entries in the same locations in A, B are added to give the entry of $A + B$ in that location.

EXAMPLE 3.3.3.

$$\begin{bmatrix} 1 & 2 & 0 \\ -1 & 3 & 1 \end{bmatrix} + \begin{bmatrix} 0 & 1 & 1 \\ 2 & 3 & -1 \end{bmatrix} = \begin{bmatrix} 1 & 3 & 1 \\ 1 & 6 & 0 \end{bmatrix}$$

Multiplication is more complicated. For each i, j to obtain the (i, j)-entry of a matrix AB multiply row i of A by column B of j. The product of a row and a column is obtained adding all products of the kth element of the row times the kth element of the column for all k. This is lengthy. If done correctly by the standard method matrix multiplication requires n^3 multiplications (n per entry of AB) and $n^3 - n^2$ additions ($n - 1$ per entry of AB) for two $n \times n$ matrices.

EXAMPLE 3.3.4.

$$\begin{bmatrix} 1 & 2 & 0 \\ -1 & 3 & 1 \\ 4 & 2 & 2 \end{bmatrix} \begin{bmatrix} 0 & 1 & 1 \\ 2 & 3 & -1 \\ 0 & 0 & 1 \end{bmatrix} = \begin{bmatrix} 4 & 7 & -1 \\ 6 & 8 & -3 \\ 4 & 10 & 4 \end{bmatrix}$$

Addition is defined only if both matrices have the same number of rows and both have the same number of columns. Multiplication of an $n \times m$ matrix times an $r \times s$ matrix is possible only if $m = r$. The answer is then $n \times s$. Matrices obey the following laws, assuming addition and multiplication in S have the corresponding properties (e.g. if S is a field).

$$A + B = B + A, \quad A + (B + C) = (A + B) + C, \quad (AB)C = A(BC)$$

$$A(B + C) = AB + AC, \quad (B + C)A = BA + CA, \quad 0 + A = A$$

$$IA = AI = A$$

DEFINITION 3.3.4. For $c \in \mathbf{F}$, cA is the matrix (ca_{ij}).

EXAMPLE 3.3.5.

$$2 \begin{bmatrix} 1 & 2 \\ 0 & 5 \end{bmatrix} = \begin{bmatrix} 2 & 4 \\ 0 & 10 \end{bmatrix}$$

If S has an element -1 then $(-1)A$ will be an ***additive inverse*** $-A$ of $A : (-1)A + A = 0$.

Existence of ***multiplicative inverses*** is not always true even over **F**.

EXAMPLE 3.3.6. Let

$$A = \begin{bmatrix} 1 & 0 \\ 1 & 0 \end{bmatrix}$$

Then A has no inverse since for any B, both rows of AB are equal since both rows of A are equal.

The ***main diagonal*** of a matrix A consists of the entries a_{ii}.

DEFINITION 3.3.5. A matrix is a *diagonal matrix* if all entries not on the main diagonal are zero.

EXAMPLE 3.3.7.

$$\begin{bmatrix} 2 & 0 \\ 0 & 3 \end{bmatrix}$$

Any diagonal matrices will commute. However, the commutative law is not true for all pairs of matrices.

The (i, j)-entry of the *mth power* A^m of a matrix is denoted $a_{ij}^{(m)}$.

The *ith row* of A is written A_{i*} and the *ith column* A_{*i}. (By replacing the asterisk with numbers we obtain the entries of the row or column.) A $1 \times n$ matrix is called a ***row vector***. An $n \times 1$ matrix is called a ***column vector***.

DEFINITION 3.3.6. The *transpose* of a matrix $A = (a_{ij})$ is $A^{\mathrm{T}} = (a_{ji})$.

That is, the (i, j)-entry of A becomes the (j, i)-entry of A^{T}. Rows become columns and columns become rows. We have $(A + B)^{\mathrm{T}} = A^{\mathrm{T}} + B^{\mathrm{T}}$, $(AB)^{\mathrm{T}} = B^{\mathrm{T}}A^{\mathrm{T}}, (A^{\mathrm{T}})^{\mathrm{T}} = A$.

EXAMPLE 3.3.8.

$$\begin{bmatrix} 1 & 2 \\ 0 & 1 \end{bmatrix}^{\mathrm{T}} = \begin{bmatrix} 1 & 0 \\ 2 & 1 \end{bmatrix}$$

DEFINITION 3.3.7. A matrix A is called *symmetric* if $A = A^{\mathrm{T}}$.

Matrices can be multiplied by row vectors or column vectors of appropriate dimension:

$$(vA)_i = (\Sigma\, v_k\, a_{ki})$$

$$(Av)_i = (\Sigma\, a_{ik}\, v_k)$$

This is matrix multiplication for $1 \times n$ or $m \times 1$ matrices, so

$$A(v + w) = Av + Aw, \quad A(Bv) = (AB)v$$

$$A(av) = a(Av), \quad Iv = v, \quad 0v = 0$$

and dual rules hold for vA.

Block and ***triangular forms*** of matrices are frequently very useful. If we partition the rows of a matrix into sets of adjacent rows and the columns of a matrix into sets of adjacent columns, the entire matrix is partitioned into blocks, each of which is a submatrix. An (i, j)-block is denoted A_{ij}. Thus

$$\begin{bmatrix} A_{11} & A_{12} & \ldots & A_{1n} \\ A_{21} & A_{22} & \ldots & A_{2n} \\ & \cdots\cdots & & \\ A_{n1} & A_{n2} & \ldots & A_{nn} \end{bmatrix}$$

EXAMPLE 3.3.9.

$$\left[\begin{array}{cc|cc} 1 & 2 & 4 & 5 \\ 0 & 0 & 1 & 0 \\ \hline 1 & 1 & 3 & 2 \\ 1 & 1 & 1 & 0 \end{array}\right]$$

This matrix has been partitioned into 4 blocks

$$A_{11} = \begin{bmatrix} 1 & 2 \\ 0 & 0 \end{bmatrix}, \quad A_{12} = \begin{bmatrix} 4 & 5 \\ 1 & 0 \end{bmatrix}$$

$$A_{21} = \begin{bmatrix} 1 & 1 \\ 1 & 1 \end{bmatrix} \quad A_{22} = \begin{bmatrix} 3 & 2 \\ 1 & 0 \end{bmatrix}$$

PROPOSITION 3.3.1. *If two matrices A, B are partitioned in the same way, their sum is the matrix of blocks* $A_{ij} + B_{ij}$. *If they are* $n \times n$ *matrices, and the columns of A are partitioned in the same say as the rows of B, their product is the matrix of blocks* $\sum_k A_{ik} B_{kj}$.

Proof. The (i, j)-entry of AB is $\sum_r a_{ir} b_{rj} = \sum_k \sum_{r \in \theta_k} a_{ir} b_{rj} = \sum_k (A_{sk} B_{kt})_{ij}$ if i, j belongs to the (s, t)-block. Thus the (s, t)-block of AB is $\sum_k A_{sk} B_{kt}$. Here θ_k denotes the kth block. □

EXAMPLE 3.3.10. Let A, B be

$$\left[\begin{array}{cc|c} 1 & 1 & 1 \\ 0 & 1 & 1 \\ \hline 0 & 0 & 1 \end{array}\right], \quad \left[\begin{array}{cc|c} 0 & 1 & 0 \\ 1 & 0 & 0 \\ \hline 1 & 0 & 1 \end{array}\right]$$

Then $A + B$, AB are

$$\left[\begin{array}{cc|c} 1 & 2 & 1 \\ 1 & 1 & 1 \\ \hline 1 & 0 & 2 \end{array}\right], \quad \left[\begin{array}{cc|c} 2 & 1 & 1 \\ 2 & 0 & 1 \\ \hline 1 & 0 & 1 \end{array}\right]$$

Note that

$$\begin{bmatrix}1 & 1\\0 & 1\end{bmatrix}\begin{bmatrix}0 & 1\\1 & 0\end{bmatrix} + \begin{bmatrix}1\\1\end{bmatrix}[1 \quad 0] = \begin{bmatrix}1 & 1\\1 & 0\end{bmatrix} + \begin{bmatrix}1 & 0\\1 & 0\end{bmatrix} = \begin{bmatrix}2 & 1\\2 & 0\end{bmatrix}$$

$$\begin{bmatrix}1 & 1\\0 & 1\end{bmatrix}\begin{bmatrix}0\\0\end{bmatrix} + \begin{bmatrix}1\\1\end{bmatrix}[1] = \begin{bmatrix}1\\1\end{bmatrix}$$

$$[0 \quad 0]\begin{bmatrix}0 & 1\\1 & 0\end{bmatrix} + [1]\,[1 \quad 0] = [1 \quad 0]$$

$$[0 \quad 0]\begin{bmatrix}0\\0\end{bmatrix} + [1]\,[1] = [1]$$

DEFINITION 3.3.8. A matrix A is in *upper* (*lower*) *triangular form* if $a_{ij} = 0$ for $i > j$ $(i < j)$.

EXAMPLE 3.3.11. This matrix is in upper triangular form.

$$\begin{bmatrix}1 & 1 & 3\\0 & 1 & 2\\0 & 0 & 4\end{bmatrix}$$

PROPOSITION 3.3.2. *If two matrices are both in upper (lower) triangular form, so is their product. The diagonal entries of the product are products of the diagonal entries of the two matrices.*

Proof. Assume A, B are in upper triangular form. Suppose $i > j$. Then in $a_{ik}\, b_{kj}$ either $i > k$ or $k > j$. So $\Sigma\, a_{ik}\, b_{kj} = 0$. Suppose $i = j$. If $i \leqslant k$ and $k \leqslant j$ then $i = k = j$. So $a_{ii}\, b_{ii}$ is the only nonzero term in $\Sigma\, a_{ik}\, b_{kj}$. A similar proof holds for the lower triangular case. □

We can also combine these two concepts by considering matrices whose blocks are in triangular form. Such a matrix is said to be in *block-triangular form.*

EXERCISES

Level 1

1. Find

$$\begin{bmatrix}1 & 2\\0 & 7\end{bmatrix} + \begin{bmatrix}2 & 4\\0 & 0\end{bmatrix}$$

2. Find

$$\begin{bmatrix} 1 & 3 \\ 4 & 2 \end{bmatrix} \begin{bmatrix} 1 & 4 \\ 3 & 2 \end{bmatrix}$$

3. Find

$$2 \begin{bmatrix} 0 & 1 & -1 \\ 1 & 5 & 4 \end{bmatrix} + (-3) \begin{bmatrix} 0 & 4 & 0 \\ -5 & 0 & 0 \end{bmatrix}$$

4. Find

$$\begin{bmatrix} 1 & 4 & 0 \\ 0 & 5 & 3 \\ 2 & 2 & -5 \end{bmatrix} \begin{bmatrix} 1 & -1 & 1 \\ 2 & 1 & 1 \\ 1 & -1 & -4 \end{bmatrix}$$

5. Give an example to show AA^T may not equal A^TA.
6. Find

$$[1 \quad 1 \quad 1] \begin{bmatrix} 1 & 10 & 11 \\ 2 & -5 & -9 \\ 3 & 0 & 13 \end{bmatrix}$$

7. Assume A, B are $n \times n$ matrices. Verify that $B^{-1}A^{-1}$ is an inverse of AB by multiplying where A^{-1} is an inverse of A and B^{-1} is an inverse of B.

Level 2

1. Give an example to verify the distributive law for matrices.
2. Given an example to verify the associative law for matrices.
3. Multiply these matrices quickly using block forms and properties of the identity matrix:

$$\left[\begin{array}{cc|cc} 1 & 0 & 1 & 0 \\ 0 & 1 & 0 & 1 \\ \hline 1 & 0 & 1 & 0 \\ 0 & 1 & 0 & 1 \end{array}\right] \left[\begin{array}{cc|cc} 1 & 2 & 10 & 9 \\ 8 & 7 & 4 & 5 \\ \hline 3 & 6 & 9 & 12 \\ 11 & 9 & 7 & 5 \end{array}\right]$$

4. Show that any two matrices of this form commute

$$\begin{bmatrix} a & 0 \\ b & a \end{bmatrix}$$

Generalize this to $n \times n$ matrices.

5. Prove that if $vA = 0$ for some nonzero vector v, A cannot have an inverse A^{-1} such that $AA^{-1} = I$.
6. Find three different 3×3 matrices A such that $A^2 = I$.
7. Find a 2×2 matrix A such that $A^2 = 0$.

Level 3

1. Multiply two matrices having block form

$$\begin{bmatrix} A & B \\ 0 & D \end{bmatrix}$$

2. Give a formula for an inverse of such a matrix in terms of A^{-1}, B^{-1} by solving

$$\begin{bmatrix} A & B \\ 0 & D \end{bmatrix} \begin{bmatrix} X & Y \\ 0 & Z \end{bmatrix} = \begin{bmatrix} I & 0 \\ 0 & I \end{bmatrix}$$

3. Do the same for upper triangular block form.
4. Factor

$$\begin{bmatrix} A & B \\ C & D \end{bmatrix}$$

by upper and lower triangular block matrices, if A is invertible. Find X, Y such that

$$\begin{bmatrix} A & B \\ C & D \end{bmatrix} = \begin{bmatrix} A & 0 \\ X & I \end{bmatrix} \begin{bmatrix} I & Y \\ 0 & D - XY \end{bmatrix}$$

In principle this and the preceding exercises give an inverse for any 2×2 matrix in block form.

5. Prove $A + A^T$, AA^T, and A^TA are symmetric using the laws regarding transposes given in the text.
6. Prove for symmetric matrices A, B that AB is symmetric if and only if A, B commute.
7. Prove that a matrix A commutes with any polynomial $c_0 I + c_1 A + \ldots + c_n A_n$ where $c_i \in \mathbf{F}$.

3.4 LINEAR TRANSFORMATIONS

A ***linear transformation*** is a homomorphism of vector spaces, and is a linear function from one vector space to another. Every linear transformation can be represented by a matrix for a given basis. Conversely every matrix gives a linear transformation.

DEFINITION 3.4.1. Let **V** and **W** be the vector spaces over **F**. A function g from **V** to **W** is a *linear transformation* if and only if $g(ax + by) = ag(x) + bg(y)$ for all $a, b \in \mathbf{F}$, $x, y \in \mathbf{V}$.

EXAMPLE 3.4.1. Let **V** be the vector space of all polynomials with coefficients in a **F** containing the rationals. The following are linear transformations on **V**.

$$g(x) \to 2g(x), \quad g(x) \to \int_0^x g(x)\,dx$$
$$g(x) \to xg(x),$$
$$g(x) \to g'(x), \quad g(x) \to g(x) - g(0)$$

EXAMPLE 3.4.2. Any $n \times m$ matrix over **F** gives a linear transformation on a vector space of row vectors defined by $v \to vA$ and on a vector space of column vectors by $v \to Av$.

In fact every linear transformation between finite dimensional vector spaces can be regarded as multiplication by some matrix. If we allow infinite matrices, this is also true for infinite dimensional vector spaces.

PROPOSITION 3.4.1. *Let g be a linear transformation from a vector space* **V** *over* **F** *to a vector space* **W** *over* **F**. *Let $\{v_1, v_2, \ldots, v_n\}$ be a basis for* **V** *and let $\{w_1, w_2, \ldots, w_m\}$ be a basis for* **W**. *Let $a_{ij} \in \mathbf{F}$ be such that $g(v_i) = \Sigma\, a_{ij} w_j$. Then $g(f_1 v_1 + f_2 v_2 + \ldots + f_n v_n) = k_1 w_1 + k_2 w_2 + \ldots + k_m w_m$ where $[f_1 f_2 \ldots f_n]A = [k_1 k_2 \ldots k_m]$.*

Proof. Since

$$g(\sum_i f_i v_i) = \sum_i f_i g(v_i) = \sum_i f_i \sum_i a_{ij} w_j = \sum_j w_j \sum_i f_i a_{ij}$$

so

$$k_j = \sum_i f_i a_{ij}$$

The same formula arises from matrix multiplication. □

This result gives a 1-1 correspondence between matrices and linear transformations for any basis.

DEFINITION 3.4.2. The *image space* of a linear transformation $g : \mathbf{V} \to \mathbf{W}$ is $\{w \in \mathbf{W} : w = g(v)$ for some $v \in \mathbf{V}\}$. The *kernel,* or *null space* of g is $\{v \in \mathbf{V} : f(v) = 0\}$. The *rank* of g is the dimension of the image space.

EXAMPLE 3.4.3. Let g be the linear transformation of multiplication by

$$\begin{bmatrix} 1 & 1 \\ 2 & 2 \end{bmatrix}$$

on row vectors. Then the image space of g is the set of vectors having the form (x, x). The kernel of g is the set of vectors having the form $(-2x, x)$. The rank of g is 1.

PROPOSITION 3.4.2. *Let g be a linear transformation from a finite dimensional vector space* $\mathbf{V}$ *into a finite dimensional vector space* $\mathbf{W}$. *Then*

$$\dim\,(\text{null space}) + \dim\,(\text{image space}) = \dim \mathbf{V}$$

Proof. By Theorem 3.2.6, the null space $\mathbf{N}$ has a complement $\mathbf{W}$. Since $\mathbf{N} \oplus \mathbf{W} \simeq \mathbf{V}$, $\dim \mathbf{N} + \dim \mathbf{W} = \dim \mathbf{V}$. We will show g is an isomorphism from $\mathbf{W}$ into the image space $\mathbf{M}$. Let $z \in \mathbf{M}$. Then $z = g(y)$ for $y \in \mathbf{V}$. And $y = w + n$, $w \in \mathbf{W}$, $n \in \mathbf{N}$. So $z = g(w+n) = g(w) + g(n) = g(w) + 0 = g(w)$. So g is onto.

Suppose $g(w_1) = g(w_2)$. Then $g(w_1) - g(w_2) = 0$. So $g(w_1 - w_2) = 0$. So $w_1 - w_2 \in \mathbf{N}$. So $w_1 - w_2 \in \mathbf{N} \cap \mathbf{W} = \{0\}$. So $w_1 = w_2$. This proves g is one-to-one. So $\mathbf{W} \simeq \mathbf{M}$. So $\dim \mathbf{W} = \dim \mathbf{M}$. □

DEFINITION 3.4.3. The *row* (*column*) *space* of a matrix is the vector space spanned by its rows (columns). The *row* (*column*) *rank* is the dimension of the row (column) space. The row (column) rank of A will be denoted by $\rho_r(A)$ $(\rho_c(A))$.

EXAMPLE 3.4.4. The row and column spaces of

$$\begin{bmatrix} 1 & 1 \\ 2 & 2 \end{bmatrix}$$

are $\{(x, x) : x \in \mathbf{R}\}$ and $\{(x, 2x) : x \in \mathbf{R}\}$. The row rank and the column rank are both 1.

Note that the row (column) rank can more simply be expressed as the maximum number of independent rows (columns).

PROPOSITION 3.4.3. *Let v be regarded as a row vector. Then $vA = \Sigma\, v_i A_{i*}$. If v is regarded as a column vector, $Av = \Sigma\, A_{*j} v_j$. The row (column) space of A is equal to the image space of A on row (column) vectors.*

Proof. The j component of vA is $\Sigma\, v_i a_{ij}$, which equals the j component of $\Sigma\, v_i A_{i*}$. Thus if e_i denotes a vector with a 1 in place i and zeros elsewhere,

$e_i A = A_{i*}$. So the row space is contained in the image space. Yet if w belongs to the image space then $w = vA = \Sigma\, v_i A_{i*}$ which is contained in the row space. Therefore the image space is contained in the row space.

The proofs for column vectors are similar. □

THEOREM 3.4.4. *The row and column rank of a matrix are equal.*

Proof. Let A be an $n \times m$ matrix of column rank r. Then there are r columns of A which form a basis for the column space. Rearranging the columns will not affect the row rank or the column rank, so we will assume that $A_{*1}, A_{*2}, \ldots, A_{*r}$ form a column basis. Then if we restrict all row vectors to their first r components, we assert that this gives an isomorphism on the row space. Let p_{ij} be such that $A_{*j} = \Sigma\, A_{*i}\, p_{ij}$ where $i = 1, 2, \ldots, r$ and $j = r + 1, r + 2, \ldots, m$. Such elements exist since $A_{*1}, A_{*2}, \ldots, A_{*r}$ is a basis.

Then for any row vector A_{i*} we have

$$a_{ij} = \sum_{k=1}^{r} a_{ik}\, p_{kj}$$

by taking the i component of the equation above. This implies that for any linear combination v of the A_{i*},

$$v_j = \sum_{k=1}^{r} v_k\, p_{kj}$$

where $j = r + 1, r + 2, \ldots, m$. Thus if two vectors have identical components $v_1, v_2, \ldots, v_r$ the components $v_{r+1}, v_{r+2}, \ldots, v_m$ are also identical. This proves the assertion.

So the row space is isomorphic to a subspace of an r-dimensional space. So it has dimension at most r. So $\rho_c(A) \leqslant \rho_r(A)$. By symmetry, $\rho_r(A) \leqslant \rho_c(A)$ where $\rho_r(A)\,(\rho_c(A))$ denotes the row (column) rank of A. This proves the theorem. □

EXAMPLE 3.4.5. Let

$$A = \begin{bmatrix} 1 & 0 & 1 \\ 0 & 1 & 1 \\ 2 & 3 & 5 \end{bmatrix}$$

Then A_{*1}, A_{*2} form a basis. There is a 1–1 map on the row space taking [1 0 1] to [1 0], [0 1 1] to [0 1], [2 3 5] to [2 3]. Thus the row space is a subspace of $\mathbf{R}^2$ and is two-dimensional.

Thm $V_1 \cong V_2 \Leftrightarrow \dim V_1 = \dim V_2$

EXERCISES

Level 1

1. Find matrices to represent these linear transformations, according to Proposition 3.4.1: (a) $f(x, y) = (x + y, y)$, (b), $f(x, y, z) = (y, z, x)$, (c) $f(x, y, z) = x + y + z$.
2. If the rows of a matrix are independent, what is its rank? What is the rank of

$$\begin{bmatrix} 1 & 1 & 1 \\ 0 & 1 & 1 \\ 0 & 0 & 1 \end{bmatrix}?$$

4. Show the zero matrix is the matrix of the linear transformation $f(x) = 0$.
5. Show if A has rank k, AB and BA have rank $\leqslant k$ for any B.

Level 2

1. Find matrices to represent these linear transformations on the space of polynomials of degree $\leqslant 5$ with basis $1, x, x^2, x^3, x^4$: (a) $f(x) - f(0)$, (b) $f'(x)$, (c) $\frac{1}{x}\int_0^x f(x)\,\mathrm{d}x$, (d) $(xf(x))'$.
2. What are the ranks of the matrices in Exercise 1?
3. Prove linear transformations from $\mathbf{V}$ to $\mathbf{W}$ form a vector space. If $\mathbf{V}$ has dimension n and $\mathbf{W}$ has dimension m, what is the dimension of the space of linear transformations from $\mathbf{V}$ to $\mathbf{W}$?
4. Show the identity matrix is the matrix of the identity linear transformation.

Level 3

1. For any vector $v \in \mathbf{R^n}$ show the transformations $x \rightarrow (x \cdot v)v$ and $x \rightarrow x - (x \cdot v)v$ are linear where $x \cdot v$ is the inner product $\Sigma\, x_i v_i$. What are the kernels?
2. Identify the ranks and image spaces in the above exercise, and describe these mappings geometrically. Prove that $f \circ f = f$ for these linear transformations.
3. Prove composition of linear transformations obeys the distributive law on each side $f(g + h) = fg + fh$, $(g + h)f = gf + hf$.
4. Find a general form for rank 1 matrices.
5. Prove that if a multiple of one row is added to another row, the rank of a matrix is unchanged. Represent such a row operation as multiplication by a matrix $I + cE(i, j)$ where c times row i is added to row j and $E(i, j)$ has a 1 entry in location i, j and all other entries are 0.
6. Describe the infinite matrices of $f \rightarrow xf$ and $f \rightarrow f'$ on the space of all polynomials of any degree with basis $1, x, x^2, \ldots, x^n, \ldots$. Show these matrices obey the identity $AB - BA = I$, used in matrix mechanics.
7. Prove any two vector spaces of the same dimension are isomorphic.

8. Prove any $n \times m$ rank k matrix can be represented as AIB where A is $n \times k$, I is a $k \times k$ identity matrix, and B is $k \times m$.

3.5 DETERMINANTS AND CHARACTERISTIC POLYNOMIALS

The determinant of an $n \times n$ matrix gives a formula for solving linear equations. However, it is more useful for theoretical purposes, and for finding the eigenvalues of a matrix. In particular it gives a criterion for a matrix to be of rank n, and to have an inverse.

DEFINITION 3.5.1. Let σ be a permutation on $\{1, 2, \ldots, n\}$, i.e. a 1-1 function from this set to itself. Then the number of inversions of σ is $|\{(i, j) : i < j \text{ and } \sigma(i) > \sigma(j)\}|$. The permutation σ is said to be *even* or *odd* according to this number being even or odd. We write sign $(\sigma) = +1$ or -1 as σ is even or odd.

We write permutations in the following form:

$$\begin{pmatrix} 1 & 2 & 3 \\ 1 & 3 & 2 \end{pmatrix}$$

This means $\sigma(1) = 1$, $\sigma(2) = 3$, $\sigma(3) = 2$. In general $\sigma(i)$ is the number underneath i.

EXAMPLE 3.5.1. The permutation

$$\begin{pmatrix} 1 & 2 & 3 \\ 2 & 3 & 1 \end{pmatrix}$$

inverts the pairs (1, 2) and (2, 3). So it is even. The permutation

$$\begin{pmatrix} 1 & 2 & 3 \\ 1 & 3 & 2 \end{pmatrix}$$

inverts the pair (2, 3). So it is odd.

PROPOSITION 3.5.1. *sign* $(\sigma_1\sigma_2) =$ *sign* (σ_1) *sign* (σ_2). *If* σ *interchanges exactly two elements then sign* $(\sigma) = -1$.

Proof. A pair (i, j) is inverted by $\sigma_1\sigma_2$ if and only if it lies in one of the sets $S = \{(i, j) : \sigma_2 \text{ inverts } (i, j)\}$ or $T = \{(i, j) : \sigma_1 \text{ inverts } (\sigma_2(i), \sigma_2(j))\}$, but not both. The cardinality of this set equals $|S \cup T| - |S \cap T| = |S| + |T| - 2|S \cap T|$ which is an even number plus $|S| + |T|$. Thus sign $(\sigma_1\sigma_2) = (-1)^{|S|+|T|}$, sign $(\sigma_2) = (-1)^{|S|}$, sign $(\sigma_1) = (-1)^{|T|}$. This proves the first assertion.

For the second assertion, let σ interchange i, j. Then if $i < k < j$ both (i, k) and (k, j) are inverted. One other pair is inverted: (i, j) itself. So σ is odd. $\square$

DEFINITION 3.5.2. The *determinant* of an $n \times n$ matrix A is

$$\det(A) = \sum_{\sigma} \operatorname{sign}(\sigma)\, a_{1\sigma(1)}\, a_{2\sigma(2)} \cdots a_{n\sigma(n)}$$

where the summation ranges over all permutations σ of $\{1, 2, \ldots, n\}$.

EXAMPLE 3.5.2.

$$\det \begin{bmatrix} a & b & c \\ d & e & f \\ g & h & i \end{bmatrix} = aei + bfg + cdh - ceg - fha - dbi$$

A number of results about determinants are important. These will be stated in an order in which it is convenient to prove them, but without proofs. Proofs can be found in most books on the subject.

D1. A determinant changes sign if two rows or two columns of a matrix are interchanged.

D2. A determinant is linear as a function of the ith column or the jth row.

D3. A determinant of a diagonal matrix is the product of the main diagonal entries.

D4. If any row or any column of a matrix is zero, then the determinant is zero.

D5. If the matrix B is obtained from A by replacing A_{j*} by $A_{j*} - kA_{i*}$ where $i \neq j$, then $\det(B) = \det(A)$. Likewise for columns.

D6. $\det(A) = \det(A^T)$.

D7. If B is obtained from A by rearranging the rows by a permutation σ then $\det(B) = \operatorname{sign}(\sigma)\det(A)$.

D8. $\det(AB) = \det(A)\det(B)$.

D9. $\det(A) \neq 0$ if and only if A has an inverse A^{-1} such that $AA^{-1} = A^{-1}A = I$.

D10. Let $A[i\,|\,j]$ be the $(n-1) \times (n-1)$ matrix obtained from A by removing the ith row and the jth column. Let $C[i\,|\,j]$ be the (i, j)th-cofactor of $A[i\,|\,j]$ which is $(-1)^{i+j}\det(A[i\,|\,j])$. Then

$$\det(A) = \sum_{j=1}^{n} a_{rj}\, C[r\,|\,j] = \sum_{i=1}^{n} a_{is}\, C[i\,|\,s]$$

for any, r, s.

D11. Suppose A has the form

$$\begin{bmatrix} A_{11} & 0 & \ldots & 0 \\ A_{21} & A_{22} & \ldots & 0 \\ & \cdots\cdots & & \\ A_{r1} & A_{r2} & \ldots & A_{rr} \end{bmatrix}$$

where the zeros and A_{ij} represent submatrices (blocks) and not single entries, and the A_{ii} are square matrices. Then

$$det\,(A) \;=\; det\,(A_{11})\; det\,(A_{22}) \ldots det\,(A_{rr})$$

D12. For any matrix A, let cof (A) be the matrix $C[\,j\,|\,i\,]$. Then $A\,(cof\,(A)) = (cof\,(A))\,A = I\,(det\,(A))$. Thus if A has an inverse,

$$A^{-1} \;=\; \frac{cof\,(A)}{det\,(A)}$$

The matrix cof (A) is called the adjoint of A.

EXAMPLE 3.5.3. Let A be

$$\begin{bmatrix} 1 & 1 & 1 \\ 1 & 2 & 3 \\ 1 & 4 & 9 \end{bmatrix}$$

Then cof (A) is

$$\begin{bmatrix} 6 & -5 & 1 \\ -6 & 8 & -2 \\ 2 & -3 & 1 \end{bmatrix}$$

and A(cof (A)) is

$$\begin{bmatrix} 2 & 0 & 0 \\ 0 & 2 & 0 \\ 0 & 0 & 2 \end{bmatrix}$$

The characteristic polynomial of a matrix gives properties of linear transformation which do not depend on the particular basis chosen.

DEFINITION 3.5.3. The *characteristic polynomial* of the matrix A is $\det\,(tI - A)$ where $t \in \mathbf{C}$.

EXAMPLE 3.5.4. The characteristic polynomial of the matrix A in Example 3.5.3 is $t^3 - 12t^2 + 15t - 2$, i.e. the determinant of

$$\begin{bmatrix} t-1 & -1 & -1 \\ -1 & t-1 & -3 \\ -1 & -4 & t-9 \end{bmatrix}$$

The characteristic polynomial of an $n \times n$ matrix has degree n. Its constant term is $\det(-A) = (-1)^n \det(A)$ (set $t = 0$).

DEFINITION 3.5.4. The *trace* of A of an $n \times n$ matrix A is $\operatorname{Tr}(A) = a_{11} + a_{22} + \ldots + a_{nn}$.

EXAMPLE 3.5.5. Let A be as in Example 3.5.3. Then $\operatorname{Tr}(A) = 12$.

The coefficient of t^{n-1} in the characteristic polynomial is $-\operatorname{Tr}(A)$. The trace has additive properties somewhat similar to the multiplicative properties of the determinant. For instance $\operatorname{Tr}(A + B) = \operatorname{Tr}(A) + \operatorname{Tr}(B)$, $\det(AB) = \det(A)\det(B)$.

PROPOSITION 3.5.2. *The coefficient of t^{n-1} in the characteristic polynomial is $-Tr(A)$. The constant term is $(-1)^n \det(A)$.*

Proof. By inspection of $(tI - A)$, the coefficient of t^{n-1} is the sum of $-a_{ii}$. Set $t = 0$. Then $\det(tI - A)$ becomes $\det(-A) = (-1)^n \det(A)$. □

THEOREM 3.5.3. (Cayley-Hamilton.) *Let $p(t)$ be the characteristic polynomial of A. Then $p(A) = 0$.*

Proof. Let $B(t) = (tI - A)$ and let $\bar{B}(t) = \bar{B}_0 + \bar{B}_1 t + \bar{B}_2 t^2 + \ldots + \bar{B}_{n-1} t^{n-1}$ be the adjoint of B (to expand $\bar{B}(t)$, let $\bar{B}_i$ be the matrix of coefficients of t^{n-1} of t^{n-1} in $\bar{B}(t)$). Then $B(t)\bar{B}(t) = \det(B(t))I = p(t)I$. So

$$
\begin{aligned}
(tI - A)(\bar{B}_0 + \bar{B}_1 t + \bar{B}_2 t^2 + \ldots + \bar{B}_{n-1} t^{n-1}) \\
&= -A\bar{B}_0 + t(\bar{B}_0 - A\bar{B}_1) + \ldots + t^n \bar{B}_{n-1} \\
&= p(t)I
\end{aligned}
$$

This equation must be an identity in t. So if

$$p(t) = t^n + c_{n-1}t^{n-1} + \ldots + c_0,\ c_i I = \bar{B}_{i-1} - A\bar{B}_i$$

We expand

$$
\begin{aligned}
(AI - A)(\bar{B}_0 + A\bar{B}_1 + \ldots + A^{n-1}\bar{B}_{n-1}) \\
&= -A\bar{B}_0 + A(\bar{B}_0 - A\bar{B}_1) + \ldots + A^n \bar{B}_{n-1} \\
&= c_0 I + c_1 A + c_2 A^2 + \ldots + A^n \\
&= p(A)
\end{aligned}
$$

But $AI - A = A - A = 0$. So $p(A) = 0$. □

EXERCISES

Level 1

1. Compute the determinant of

$$\begin{bmatrix} 1 & 2 \\ 2 & 3 \end{bmatrix}, \begin{bmatrix} 1 & 1 & 1 \\ 1 & 2 & 3 \\ 1 & 4 & 9 \end{bmatrix}, \begin{bmatrix} 1 & 1 & 0 \\ 0 & 1 & 1 \\ 1 & 0 & 1 \end{bmatrix}$$

2. Compute the characteristic polynomial of

$$\begin{bmatrix} 0 & 1 \\ -1 & 0 \end{bmatrix}, \begin{bmatrix} 0 & 1 & 0 \\ 0 & 0 & 1 \\ 0 & 0 & 0 \end{bmatrix}$$

3. Prove Property D3 of determinants.
4. Is this permutation (123) (45) even or odd?
5. Prove Property D4 of determinants.

Level 2

1. Compute the determinant of

$$\begin{bmatrix} a & b & c \\ c & a & b \\ b & c & a \end{bmatrix}, \begin{bmatrix} 0 & a & b \\ -a & 0 & c \\ -b & -c & 0 \end{bmatrix}$$

2. Prove Property D2 of determinants. Show also that if all entries are multiplied by a constant c the determinant is multiplied by c^n where n is the order of the matrix.
3. Use Exercise 2 and Property D6 to show that the determinant of an $n \times n$ matrix A is zero if $A^T = -A$ and n is odd.
4. Prove that if two rows of a determinant are equal its value is zero.
5. Prove Property D5 of determinants.
6. Use Property D11 to compute the determinant of

$$\begin{bmatrix} 1 & 2 & 0 & 0 \\ 2 & 1 & 0 & 0 \\ 10 & 11 & 3 & 5 \\ 14 & 15 & 5 & 3 \end{bmatrix}$$

7. Find the inverse of

$$\begin{bmatrix} 1 & 1 & 1 \\ 1 & 2 & 4 \\ 1 & 4 & 16 \end{bmatrix}$$

Level 3

1. Prove Property D7 of determinants.
2. Prove Property D8 of determinants. Show $\det(AB)$ is a function which is (i) linear in each row, (ii) if two rows of A are equal then it is zero. Thus it is unchanged under row operations as in Level 2, Exercise 5. Prove under permutation of the rows of A we have a property like D7. Now by such operations we can reduce A to a matrix having at most one nonzero entry per column. Prove directly $\det(AB) = \det(A)\det(B)$ in this case.
3. Find a general formula for the inverse of a 3×3 matrix.
4. Find recursion formulas for the characteristic polynomial of an $n \times n$ $(0,1)$-matrix A such that $a_{ij} = 1$ if $i \geqslant j - 1$, for example

$$\begin{bmatrix} 1 & 1 & 0 & 0 \\ 1 & 1 & 1 & 0 \\ 1 & 1 & 1 & 1 \\ 1 & 1 & 1 & 1 \end{bmatrix}$$

Let f_n, g_n be the determinants of the $n \times n$ matrices corresponding to

$$\begin{bmatrix} x & 1 & 0 & 0 \\ 1 & x & 1 & 0 \\ 1 & 1 & x & 1 \\ 1 & 1 & 1 & x \end{bmatrix}, \begin{bmatrix} 1 & 1 & 0 & 0 \\ 1 & x & 1 & 0 \\ 1 & 1 & x & 1 \\ 1 & 1 & 1 & x \end{bmatrix}$$

Prove by Property D10 that $f_n(x) = xf_{n-1}(x) - g_{n-1}(x)$ and $g_n(x) = f_{n-1}(x) - g_{n-1}(x)$. Note $(-1)^n f_n(1-t)$ is the desired characteristic polynomial.
5. Prove using the formulas in Exercise 4 that if $h_n = (-1)^n f_n(1-t)$ it satisfies $h_n = th_{n-1} - th_{n-2}$. Also $h_1 = (t-1)$, $h_2 = t^2 - 2t$ (we assume $h_0 = 1$).
6. Prove that h_n is the coefficient of u^n in

$$\frac{1-u}{1-ut+u^2t}$$

by showing this coefficient satisfies the same recursion formula and has the same value for $n = 0, 1$.

7. Prove Property D9 for determinants by using row operations as in Exercise 2 to simplify A.
8. Show every polynomial in an $n \times n$ matrix A can be reduced to a sum of powers of $I, A, A^2, \ldots, A^{n-1}$. First express A^n this way using the Cayley-Hamilton Theorem. Then express A^{n+1} in terms of $A^n, A^{n-1}, \ldots, A$ using that expression and substitute the previous expression for A^n.

9 Find M^{100} where $M = (\quad)$ → to p. 156

3.6 EIGENVALUES, EIGENVECTORS, SIMILARITY

If we change the basis in a vector space, the matrix A representing any linear transformation will not be unchanged but will be replaced by a matrix XAX^{-1}. Thus the intrinsic properties of the transformation will be reflected by those properties of A not affected by the linear transformation. A matrix XAX^{-1} is said to be ***similar*** to A. Thus we look for similarity invariants. The most important is the characteristic polynomial and its roots, called *eigenvalues*.

DEFINITION 3.6.1. Two $n \times n$ matrices A, B are *similar* if and only if there exists an $n \times n$ matrix X, having an inverse, such that $B = XAX^{-1}$.

EXAMPLE 3.6.1. The matrices A, B

$$\begin{bmatrix} 0 & 1 \\ 1 & 0 \end{bmatrix}, \begin{bmatrix} 1 & 0 \\ 0 & -1 \end{bmatrix}$$

are similar. Let X be the matrix

$$\begin{bmatrix} 1 & 1 \\ -1 & 1 \end{bmatrix}$$

then X is invertible and $XA = BX$.

PROPOSITION 3.6.1. *For a fixed matrix X, the mapping $A \to XAX^{-1}$ is an isomorphism of rings.* — def. is not introduced (see p. 164)

Proof. We have $X(AB)X^{-1} = (XAX^{-1})(XBX^{-1})$, $X(A+B)X^{-1} = XAX^{-1} + XBX^{-1}$ and an inverse mapping is given by $X^{-1}AX$. □

PROPOSITION 3.6.2. *The relation of similarity is an equivalence relation.*

Proof. (Reflexivity) $A = IAI^{-1}$. (Symmetry) if $A = XBX^{-1}$ then $B = (X^{-1})A(X^{-1})^{-1}$. (Transitivity) if $A = XBX^{-1}$ and $B = YCY^{-1}$ then $A = (XY)C(XY)^{-1}$. □

PROPOSITION 3.6.3. *Let f be a linear transformation from* **V** *to* **V**. *Let* $\{v_1, v_2, \ldots, v_n\}$ *and* $\{w_1, w_2, \ldots, w_n\}$ *be bases for the vector space* **V**. *Let A be the matrix of f in terms of the basis* $\{v_1, v_2, \ldots, v_n\}$ *and let B be the matrix of f in terms of the basis* $\{w_1, w_2, \ldots, w_n\}$. *Let matrices X and Y be defined by expanding* $v_j = \Sigma\, w_i x_{ij}$ *and* $w_j = \Sigma\, v_i y_{ij}$. *Then* $Y = X^{-1}$ *and* $XA = BX$. *Thus A, B are similar.*

Proof. The matrix A is defined by $f(v_j) = \Sigma\, v_i a_{ij}$ and B is defined by $f(w_j) = \Sigma\, w_i b_{ij}$. Thus

$$\begin{aligned} f(v_j) &= \sum_i v_i a_{ij} = \sum_i \sum_k w_k x_{ki} a_{ij} \\ &= \sum_k w_k \sum_i x_{ki} a_{ij}. \end{aligned}$$

On the other hand

$$\begin{aligned} f(v_j) &= f(\sum_i w_i x_{ij}) = \sum_i x_{ij} f(w_i) \\ &= \sum_i x_{ij} \sum_k w_k b_{ki} = \sum_k w_k \sum_i b_{ki} x_{ij}. \end{aligned}$$

Therefore

$$\sum_k w_k \sum_i x_{ki} a_{ij} = \sum_k w_k \sum_i b_{ki} x_{ij}.$$

Since the set $\{w_1, w_2, \ldots, w_n\}$ is independent,

$$\sum_i x_{ki} a_{ij} = \sum_i b_{ki} x_{ij}.$$

So $XA = BX$. By the same sort of argument, $XY = I$. □

PROPOSITION 3.6.4. *The characteristic polynomial is invariant under similarity.*

Proof. $\det(tI - XAX^{-1}) = \det(X(tI - A)X^{-1}) = \det(X)\det(tI - A)(\det(X))^{-1} = \det(tI - A)$. □

COROLLARY 3.6.5. *All the coefficients of the characteristic polynomial, such as the trace and determinant, are also similarity invariants.*

DEFINITION 3.6.2. A nonzero row (column) vector v is said to be a *row* (*column*) *eigenvector* of the matrix A if and only if $vA = kv$ ($Av = kv$) for some $k \in \mathbf{C}$. The complex number k is called an *eigenvalue*.

EXAMPLE 3.6.2. The vectors $(1, 1)$ and $(1, -1)$ are both row and column eigenvectors of the matrix

$$\begin{bmatrix} 0 & 1 \\ 1 & 0 \end{bmatrix}$$

The corresponding eigenvalues are 1 and -1.

PROPOSITION 3.6.6. *An element $k \in \mathbf{F}$ is an eigenvalue of A if and only if k is a root of the characteristic polynomial of A.*

Proof. The element k is a (row) eigenvalue if and only if $vA = kv$ for some $v \neq 0$ if and only if $0 = v(kI - A)$ for some $v \neq 0$ if and only if $kI - A$ has a nonzero kernel if and only if $\det(kI - A) = 0$ if and only if k is a root of the characteristic polynomial. □

DEFINITION 3.6.3. The *multiplicity* of an eigenvalue k is its multiplicity as a root of the characteristic polynomial $p(t)$, i.e. the highest power of $(t - k)$ which divides $p(t)$.

EXAMPLE 3.6.3. The matrix

$$\begin{bmatrix} 1 & 0 \\ 0 & 1 \end{bmatrix}$$

has the eigenvalue 1 with multiplicity 2, since the characteristic polynomial is $t^2 - 2t + 1 = (t - 1)^2$.

PROPOSITION 3.6.7. *The trace is the sum of the eigenvalues, each counted with its multiplicity. The determinant is the product of the eigenvalues, each counted with its multiplicity.*

Proof. Let the characteristic polynomial be $x^n + c_1 x^{n-1} + \ldots + c_n$. Then by Proposition 3.5.2, the trace is $-c_1$, and the determinant is $(-1)^n c_n$. But for any polynomial of degree n, the coefficient of x^{n-1} is $-$(sum of roots) and the constant term is $(-1)^n$(product of roots) if the coefficient of x^n is 1. This proves the proposition. □

From here on, in this section, we assume that the field $\mathbf{F}$ is such that the characteristic polynomial $p(t)$ factors into linear factors $t - k_i$ over $\mathbf{F}$. For instance this will be true for any matrix of real or complex numbers if we use $\mathbf{F} = \mathbf{C}$. We will also assume vectors are row vectors unless the contrary is stated.

Let the eigenvalues of A be $k_1, k_2, \ldots, k_r$ and their multiplicities be $n_1, n_2, \ldots, n_r$. Then the characteristic polynomial is $p(t) = (t - k_1)^{n_1}(t - k_2)^{n_2} \ldots (t - k_r)^{n_r}$.

DEFINITION 3.6.4. The kernel of $(k_i I - A)^{n_i}$ is called the *characteristic subspace* belonging to k_i.

EXAMPLE 3.6.4. The matrix

$$\begin{bmatrix} 1 & 0 & 0 \\ 1 & 1 & 0 \\ 0 & 0 & -1 \end{bmatrix}$$

has two eigenvalues: 1, with multiplicity 2, and -1, with multiplicity 1. The characteristic subspace belonging to 1 is $\{(x, y, 0) : x, y \in \mathbf{R}\}$. The characteristic subspace belonging to -1 is $\{(0, 0, z) : z \in \mathbf{R}\}$.

Let V_i denote the subspace belonging to k_i.

We will use the following result about polynomials: if $f_1(t), f_2(t), \ldots, f_r(t)$ are polynomials over a field $\mathbf{F}$ having greatest common divisor 1, then there exist polynomials $a_1(t), a_2(t), \ldots, a_r(t)$ such that $1 = a_1(t)f_1(t) + a_2(t)f_2(t) + \ldots + a_r(t)f_r(t)$. We also assume unique factorization of such polynomials.

THEOREM 3.6.8. *The vector space* $\mathbf{V}$ *is the (internal) direct sum of* $V_1, V_2, \ldots, V_r$. *For* $i = 1$ *to* r, $(V_i)A \subset V_i$. *The dimension of* V_i *is* n_i. *There exists a matrix* X *such that* XAX^{-1} *has the form*

$$\begin{bmatrix} B\langle 1\rangle & 0 & \ldots & 0 \\ 0 & B\langle 2\rangle & \ldots & 0 \\ & \ldots\ldots & & \\ 0 & 0 & \ldots & B\langle r\rangle \end{bmatrix}$$

where the zeros are zero submatrices and the $B\langle i\rangle$ *are square submatrices (blocks). The characteristic polynomial of* $B\langle i\rangle$ *is* $(t - k_i)^{n_i}$.

Proof. Let $f_i(t) = p(t)(t - k_i)^{-n_i}$. Then $f_i(t)$ is a polynomial, and the different polynomials f_i have greatest common divisor 1. Thus $1 = a_1(t)f_1(t) + a_2(t)f_2(t) + \ldots + a_r(t)f_r(t)$ for some polynomials a_i. Such an identity will mean that the coefficients of each power of t are equal on both sides. This implies that the identity will remain true if t is replaced by A. Thus $I = a_1(A)f_1(A) + a_2(A)f_2(A) + \ldots + a_r(A)f_r(A)$. Multiply both sides by the row vector v. So $v = va_1(A)f_1(A) + va_2(A)f_2(A) + \ldots + va_r(A)f_r(A)$. Let $v\langle i\rangle = va_i(A)f_i(A)$. Then $v\langle i\rangle(A - k_iI)^{n_i} = va_i(A)f_i(A)(A - k_iI)^{n_i} = va_i(A)p(A) = 0$ by the Cayley-Hamilton theorem.

So $v\langle i\rangle \in V_i$. This proves that every vector is a sum of vectors lying in the subspaces V_i. So $\mathbf{V} = V_1 + V_2 + \ldots + V_r$.

We must also show that for any choice of $u\langle i\rangle \in V_i$, that if $u\langle 1\rangle + u\langle 2\rangle + \ldots + u\langle r\rangle = 0$ then $u\langle i\rangle = 0$. We have $u\langle i\rangle = u\langle i\rangle a_1(A)f_1(A) + u\langle i\rangle a_2(A)f_2(A) + \ldots + u\langle i\rangle a_r(A)f_r(A) = u\langle i\rangle f_1(A)a_1(A) + u\langle i\rangle f_2(A)a_2(A) + \ldots + u\langle i\rangle f_r(A)a_r(A)$ since any two powers of A commute. But for $j \neq i$, $f_j(A)$ has the factor $(A - k_i I)^{n_i}$ so since $u\langle i\rangle \in \ker (A - k_i I)^{n_i}$, $u\langle i\rangle f_j(A) = 0$. So $u\langle i\rangle = u\langle i\rangle f_i(A) a_i(A)$. Apply $f_i(A) a_i(A)$ to the equation

$$0 = u\langle 1\rangle + u\langle 2\rangle + \ldots + u\langle r\rangle,$$

$$0 = 0 + 0 + \ldots + u\langle i\rangle + 0 + \ldots + 0.$$

So each $u\langle i\rangle$ is 0. This proves that **V** is the direct sum of $V_1, V_2, \ldots, V_r$. If $v(A - k_i I)^{n_i} = 0$ then $vA(A - k_i I)^{n_i} = v(A - k_i I)^{n_i} A = 0A = 0$. This proves $(V_i)A \subset V_i$.

Let $d_i = \dim(V_i)$. Choose vectors $w_1, w_2, \ldots, w_n$ such that $w_1, w_2, \ldots, w_{d_1}$ is a basis for V_1, $w_{d_1+1}, w_{d_1+2}, \ldots, w_{d_1+d_2}$ is a basis for V_2, and so on. Then $w_1, w_2, \ldots, w_n$ is a basis for A. Let B be the matrix of the linear transformation given by A, with the $w_1, w_2, \ldots, w_n$ as basis. Then B is similar to A, by Proposition 3.6.5. And since $(V_i)A \subset V_i$, $w_i A$ is a linear combination of those w_j which lie in the same subspace V_i. This means that B has the form

$$\begin{bmatrix} B\langle 1\rangle & 0 & \ldots & 0 \\ 0 & B\langle 2\rangle & \ldots & 0 \\ & & \ldots\ldots & \\ 0 & 0 & \ldots & B\langle r\rangle \end{bmatrix}$$

where $B\langle i\rangle$ are certain submatrices corresponding to the subspaces V_i. For instance $B\langle 1\rangle$ is the submatrix of b_{ij} such that $i, j = 1, 2, \ldots, d_1$. Then the characteristic polynomial of B by Property D11 of determinants is the product of the characteristic polynomials of the $B\langle i\rangle$. Thus $p(t) = (t - k_1)^{n_1}(t - k_2)^{n_2} \ldots (t - k_r)^{n_r} = g_1(t) g_2(t) \ldots g_r(t)$ where g_i is the characteristic polynomial of $B\langle i\rangle$. But on V_i. $(A - k_i I)^{n_i} = 0$, by definition of V_i. Also $g_i(A) = 0$ on V_i. Suppose $g_i(t)$ has a root f other than k_i. Then f is an eigenvalue of $B\langle i\rangle$, by Proposition 3.6.6. So for some $v \in V_i$, $vA = fv$. Then $v(A - k_i I)^{n_i} = v(f - k_i)^{n_i} \neq 0$. This is a contradiction. So $g_i(t) = (t - k_i)^{d_i}$. So $(t - k_1)^{n_1} (t - k_2)^{n_2} \ldots (t - k_r)^{n_r} = (t - k_1)^{d_1}(t - k_2)^{d_2} \ldots (t - k_r)^{d_r}$ where $k_1, k_2, \ldots, k_r$ are distinct. So $n_i = d_i$. This proves the theorem. □

The next result, which we do not prove, gives the general characterization of similarity by showing that every matrix only be transformed into a unique form, the ***Jordan Canonical Form***. In this form in addition to diagonal entries the only nonzero entries possible are 1 entries immediately below the main diagonal.

EXAMPLE 3.6.5. These matrices are in Jordan Canonical Form.

$$\begin{bmatrix} 2 & 0 & 0 \\ 1 & 2 & 0 \\ 0 & 1 & 2 \end{bmatrix}, \begin{bmatrix} 2 & 0 & 0 \\ 0 & 2 & 0 \\ 0 & 1 & 2 \end{bmatrix}, \begin{bmatrix} 4 & 0 & 0 \\ 0 & 1 & 0 \\ 0 & 1 & 1 \end{bmatrix}$$

However, the eigenvalues which are the numbers appearing on the main diagonal, are sufficient for many purposes.

A weaker result, a lower triangular form, is established in the exercises. This suffices to prove the last result of this section.

THEOREM 3.6.9. (Jordan Decomposition Theorem.) *Every matrix over a field containing its eigenvalues is similar to a direct sum of matrices A such that* $a_{ij} = 0$ *unless* $i = j$ *or* $i = j + 1$, $a_{ii} = k$, *an eigenvalue of A, and* $a_{i+1,i} = 1$. *The summands are unique except for order.*

For a proof consult a more advanced book on linear algebra or matrix theory.

The following result is useful in finding the eigenvalues of many matrices.

PROPOSITION 3.6.10. *Let p be a polynomial. Then the eigenvalues of* $p(A)$ *are the elements* $p(k_i)$, *each counted with multiplicity* n_i.

Proof. By the preceding theorem, we may assume that A is a matrix whose main diagonal entries are k_i, having no nonzero entries above the main diagonal, and diagonal entries of A^s will be k_i^s. Therefore, adding the different powers of A in $p(A)$, we find that $p(A)$ has no nonzero entries above the main diagonal and the main diagonal entries are $p(k_i)$. By Property D11 of determinants, the characteristic polynomial of $p(A)$ will be the product of $(t - p(k_i))^{n_i}$. □

EXERCISES

Level 1

1. Find the eigenvalues of these matrices.

$$\begin{bmatrix} 3 & 4 \\ 1 & 3 \end{bmatrix}, \begin{bmatrix} 0 & 1 & 1 \\ -1 & 0 & 1 \\ -1 & -1 & 0 \end{bmatrix}, \begin{bmatrix} 3 & 4 & 0 \\ 4 & 2 & 0 \\ 0 & 0 & 1 \end{bmatrix}, \begin{bmatrix} 1 & 1 & 1 & 1 \\ 1 & 1 & 1 & 1 \\ 1 & 1 & 1 & 1 \\ 1 & 1 & 1 & 1 \end{bmatrix}$$

2. Any 3 × 3 *circulant*

$$\begin{bmatrix} a & b & c \\ c & a & b \\ b & c & a \end{bmatrix}$$

is $aI + bP + cP^2$, where

$$P = \begin{bmatrix} 0 & 1 & 0 \\ 0 & 0 & 1 \\ 1 & 0 & 0 \end{bmatrix}$$

Find the eigenvalues of a 3×3 circulant.

3. Any matrix

$$\begin{bmatrix} a & b & b & b \\ b & a & b & b \\ b & b & a & b \\ b & b & b & a \end{bmatrix}$$

is $(a-b)I + bJ$ where J is the matrix all of whose entries are 1. Find a formula for the eigenvalues of such a matrix.

4. What are the eigenvalues of a diagonal matrix?

Level 2

1. Show by Theorem 3.6.8 that any matrix with distinct eigenvalues is similar to a diagonal matrix.
2. If a matrix has distinct eigenvalues $t_1, t_2, \ldots, t_k$ what diagonal matrix will it be similar to?
3. Prove by induction that any matrix over **C** is similar to a lower triangular matrix. Let v_1 be an eigenvector and W a complementary subspace with basis $v_2, v_3, \ldots, v_n$. Express the linear transformation in terms of $v_1, v_2, \ldots, v_n$. Show there is a block triangular form. If the matrix has already been made similar to a lower triangular matrix on W (by choosing a suitable basis) this completes the proof. This suffices to establish the last proposition above.
4. A matrix A is called ***nilpotent*** if $A^k = 0$ for some k. Show all eigenvalues of a nilpotent matrix are zero.
5. Conclude from Exercises 3, 4 that a nilpotent matrix is similar to a matrix with zeros on or above the main diagonal. Is any such matrix nilpotent?
6. If $A^2 = A$ what are the only numbers which can be eigenvalues of A?
7. Show that a matrix A has an inverse if and only if 0 is not an eigenvalue.
8. Find the eigenvalues of the $n \times n$ matrix J all of whose entries are 1. A complete set of eigenvectors is given by $(1, 1, \ldots, 1)$, $(1, -1, -1, \ldots, 0)$, $(1, 0, -1, \ldots, 0), \ldots, (1, 0, 0, \ldots, -1)$. Generalize Exercise 7, Level 1 to $n \times n$ matrices.
9. Generalize Exercises 5, 6, Level 1 to $n \times n$ matrices.

Level 3

1. Conclude from the Jordan Decomposition Theorem that every matrix can be expressed as a sum $A = D + N$ where $DN = ND$ and N is nilpotent and D is similar to a diagonal matrix.
2. Show that the characteristic polynomial of a matrix can be calculated from the n numbers $\mathrm{Tr}(A^k)$, $k = 1$ to n. Let the eigenvalues be x_k. Then $\mathrm{Tr}(A^k) = \Sigma\, x_k{}^k$. Let this be denoted s_k. Let c_k be the coefficients in the characteristic polynomial. They are except for sign

$$-c_1 = x_1 + x_2 + \ldots + x_n$$

$$c_2 = \sum_{i<j} x_i x_j$$

$$-c_3 = \sum_{i<j<m} x_i x_j x_m$$

$$\ldots\ldots$$

$$(-1)^n c_n = x_1 x_2 \ldots x_m.$$

Prove Newton's formulas $kc_m + c_{m-1}s_1 + c_{m-2}s_2 + \ldots + c_0 s_m = 0$ where $c_0 = 1$. Using these, we can find in turn $c_1 = -s_1$, (set $m = 1$), $c_2 = -s_2$ (set $m = 2$), and so on. Thus we can find the characteristic polynomial of A. Its roots will be the eigenvalues.
3. Show that the eigenvalues of the matix

$$\begin{bmatrix} 1 & 1 & 0 & 0 \\ 1 & 1 & 1 & 0 \\ 1 & 1 & 1 & 1 \\ 1 & 1 & 1 & 1 \end{bmatrix}$$

considered in the exercises of the last section are in the $n \times n$ case

$$4\cos^2 \frac{2k\pi}{2n+4}.$$

Use the results of those exercises in that the characteristic polynomial is the coefficient of u^n in

$$\frac{1-u}{1-tu+tu^2}.$$
4. Construct a matrix having arbitrary characteristic polynomial of the form

$$\begin{bmatrix} 0 & 1 & 0 \\ 0 & 0 & 1 \\ a & b & c \end{bmatrix}$$

Such a matrix is called a ***companion matrix.***

5. Prove that any nilpotent matrix is similar to a $(0,1)$-matrix all of whose 1 entries are such that $i = j + 1$. Construct a basis beginning with 0 eigenvectors as a set S_0. Each basis element must be sent to another basis element or zero.

3.7 SYMMETRIC AND UNITARY MATRICES

A matrix M is ***symmetric*** if $M = M^T$ and it is ***real orthogonal*** if $M^{-1} = M^T$. ***Hermitian*** and ***unitary matrices*** are generalizations of these concepts to complex numbers, and are defined by equations $M = (M^T)^*$ or $M^{-1} = (M^T)^*$ where * is complex conjugation. Every Hermitian or unitary matrix can be represented as UDU^{-1} where U is unitary and D is diagonal.

Both unitary and Hermitian matrices are special cases of a slightly more general class called ***normal matrices***.

EXAMPLE 3.7.1. This matrix is unitary for any θ:

$$\begin{bmatrix} \cos\theta & -\sin\theta \\ \sin\theta & \cos\theta \end{bmatrix}$$

The inverse and transpose is

$$\begin{bmatrix} \cos\theta & \sin\theta \\ -\sin\theta & \cos\theta \end{bmatrix}$$

EXAMPLE 3.7.2. This matrix is Hermitian:

$$\begin{bmatrix} 2 & 2+3i \\ 2-3i & 5 \end{bmatrix}$$

Entries above the main diagonal must be complex conjugates of those below it.

DEFINITION 3.7.1. A matrix M is *normal* if and only if $M(M^T)^* = (M^T)^*M$.

EXAMPLE 3.7.3. Since $MM = MM$ and $M^{-1}M = MM^{-1} = I$ all Hermitian and unitary matrices are normal.

To deal with transposes effectively, we need the idea of an ***inner product*** of two complex row vectors.

DEFINITION 3.7.2. The *inner product* of two vectors x, y is the 1×1 matrix $x \cdot y = x(y^T)^*$. This can be written $x \cdot y = x_1 y_1^* + x_2 y_2^* + \ldots + x_n y_n^*$.

EXAMPLE 3.7.4. $(1, 2+i) \cdot (1, 2-i) = 1^2 + (2+i)(2-i) = 1+4+1=6$.

PROPOSITION 3.7.1. *Let x, y, z be any row vectors and $a \in \mathbf{F}$. The inner product has these properties:*

(*1*) *$x \cdot x$ is real and nonnegative.*
(*2*) *$x \cdot x = 0$ only if $x = 0$.*
(*3*) $x \cdot y = (y \cdot x)^*$.
(*4*) $x \cdot (y+z) = x \cdot y + x \cdot z$.
(*5*) $(y+z) \cdot x = y \cdot x + z \cdot x$.
(*6*) $ax \cdot y = a(x \cdot y)$.
(*7*) $x \cdot ay = a^*(x \cdot y)$.

Proof. Properties 1, 2 follows from

$$x \cdot x = x_1 \cdot x_1{}^* + x_2 \cdot x_2{}^* + \ldots + x_n \cdot x_n{}^* = \Sigma\, |x_i|^2$$

Property 3 follows from

$$y \cdot x = \Sigma\, y_i x_i^* = \Sigma\, (x_i y_i^*)^* = (x \cdot y)^*$$

Properties 4, 5, 6 follow from the distributive law for matrix multiplication. Property 7 follows from Properties 6 and 3. □

In general a function $b(x, y)$ with these properties is called a ***bilinear Hermitian form***.

PROPOSITION 3.7.2. *Let A, B be any matrices, and x, y be any vectors.* (*1*) *If $xAy^{\mathrm{T}} = xBy^{\mathrm{T}}$ for all x, y then $A = B$.* (*2*) $x \cdot (yA) = (x(A^{\mathrm{T}})^*) \cdot y$.

Proof. Let x, y have a 1 in places i, j respectively and zeros elsewhere. Then $xAy^{\mathrm{T}} = a_{ij} = xBy^{\mathrm{T}} = b_{ij}$. This proves (1). For (2), $x((yA)^{\mathrm{T}})^* = x(A^{\mathrm{T}})^*(y^{\mathrm{T}})^*$. □

These results are useful in proving results concerning normal matrices.

Two vectors are called ***orthogonal*** if $x \cdot y = 0$. In the case of real vectors, this means they are perpendicular to one another.

DEFINITION 3.7.3. A set S of vectors is *orthonormal* if and only if (1) $v \cdot v = 1$ for all $v \in S$, (2) for v, w distinct in S, $v \cdot w = 0$.

The first condition is called *normality* (length is 1), the second *orthogonality* (different vectors perpendicular).

EXAMPLE 3.7.5. The vectors $(1, 0, \ldots, 0), (0, 1, \ldots, 0), \ldots, (0, 0, \ldots, 1)$ are orthonormal.

Orthonormality means that the vectors can in effect be chosen as the basis of a coordinate system geometrically equivalent to standard coordinates. Orthonormal vectors must be linearly independent: if $a_1 x_1 + a_2 x_2 + \ldots + a_n x_n = 0$,

$a_1 \neq 0$, then $a_1 x_1 \cdot x_1 + a_2 x_1 \cdot x_2 + \ldots + a_n x_1 \cdot x_n = a_1 x_1 \cdot x_1 = 0$. But this is false.

LEMMA 3.7.3. *Let v be an eigenvector of a normal matrix N with eigenvalue k. Then v is an eigenvector of $(N^{\mathrm{T}})^*$ with eigenvalue k^*.*

Proof. Let $(N^{\mathrm{T}})^* = M$. It suffices to show

$$(vM - k^*v) \cdot (vM - k^*v) = 0$$

This is

$$\begin{aligned} & vM \cdot vM - k^*v \cdot vM - vM \cdot k^*v + k^*v \cdot k^*v \\ &= vMN \cdot v - k^*(vN \cdot v) - k(v \cdot Nv) + k^*kv \cdot v \\ &= vNM \cdot v - k^*(kv \cdot v) - kv \cdot kv + k^*kv \cdot v \\ &= vN \cdot vN - k^*kv \cdot v - k^*kv \cdot v + k^*kv \cdot v \\ &= kv \cdot kv - k^*kv \cdot v = kk^*v \cdot v - k^*kv \cdot v = 0 \end{aligned}$$

□

THEOREM 3.7.4. *Let W be any subspace of $\mathbf{C}^{\mathbf{n}}$. Then W has an orthonormal basis, and this basis extends for an orthonormal basis for all of $\mathbf{C}^{\mathbf{n}}$. Here $\mathbf{C}^{\mathbf{n}} = \mathbf{C} \oplus \mathbf{C} \oplus \ldots \oplus \mathbf{C}$.*

Proof. We construct an orthonormal basis of row vectors by the Gram-Schmidt Process. Let $v_1, v_2, \ldots, v_k$ be any basis for W. Take

$$w_1 = \frac{v_1}{\sqrt{v_1 \cdot v_1}}$$

We have $w_1 \cdot w_1 = 1$. Given $w_1, w_2, \ldots, w_i$ construct w_{i+1} as follows. Let $u_{i+1} = v_{i+1} - \Sigma\,(v_{i+1} \cdot w_i)w_i$. Then $u_{i+1} \cdot w_j = 0$ for $j = 1$ to i and $u_{i+1} \neq 0$. Let

$$w_{i+1} = \frac{u_{i+1}}{\sqrt{u_{i+1} \cdot u_{i+1}}}$$

Then $w_i \cdot w_i = 1$ and $w_{i+1} \cdot w_j = 0$ for $j = 1$ to i. Moreover w_{i+1} is linearly independent from $w_1, w_2, \ldots, w_i$. Therefore $w_1, w_2, \ldots, w_n$ form an orthonormal basis.

We can extend this to a basis for all of $\mathbf{C}^{\mathbf{n}}$ by extending the basis $\{v_i\}$ to a basis for all of $\mathbf{C}^{\mathbf{n}}$ and using the same process. □

LEMMA 3.7.5. *Let M be a normal matrix. Write the linear transformation $v \to vM$ in terms of a new orthonormal basis $w_1, w_2, \ldots, w_n$ to obtain a matrix X. Then X is normal.*

Proof. We have $UMU^{-1} = X$ where U is the matrix whose ith row is w_i. And by orthonormality $U(U^T)^* = I$. Therefore U is unitary. So $M(M^T)^* = (U^{-1}XU)(U^{-1}XU)^{T*} = (U^{-1}XU)(U^{T*}X^{T*}(U^{-1})^{T*}) = (U^{-1}XU)(U^{-1}X^{T*}U) = U^{-1}XX^{T*}U = U^{-1}X^{T*}XU = (U^{-1}X^{T*}U)(U^{-1}XU) = M^{T*}M$. □

Theorem 3.7.6. *Let N be a normal matrix. There exists an orthonormal basis consisting of eigenvectors of N.*

Proof. There exists an eigenvector since the characteristic polynomial has at least one root. Multiply this eigenvector by a positive real number r to obtain an eigenvector v with eigenvalue k for some k such that $v \cdot v = 1$. Let $\mathbf{W}$ be the space of vectors orthogonal to v_1 that is $\mathbf{W} = \{w : v \cdot w = 0\}$. Then for $w \in \mathbf{W}$, $v \cdot wN = v(N^T)^* \cdot w = \lambda^* v \cdot w = 0$ by Lemma 3.7.3 where $\lambda \in \mathbf{C}$. Thus $(\mathbf{W})N \subset \mathbf{W}$.

Now repeat the process restricted to $\mathbf{W}$. Find an eigenvector, multiply it by a real number and take its perpendicular space in $\mathbf{W}$. Repetition of this process constructs the required basis. (By Theorem 3.7.4 and Lemma 3.7.5 we may still take a normal matrix when the linear transformation is restricted to $\mathbf{W}$.) □

COROLLARY 3.7.7. *Let X be normal. Let U be a matrix whose rows are an orthonormal basis of eigenvectors of X and D a diagonal matrix such that the diagonal entries are the corresponding eigenvalues. Then U is unitary and $X = U^{-1}DU$.*

Proof. From orthonormality we have $U(U^T)^* = I$. Thus $U^{-1} = (U^T)^*$. And $UX = DU$ by definition of eigenvector. □

COROLLARY 3.7.8. *The eigenvalues of a Hermitian matrix are real and those of a unitary matrix have absolute value 1.*

Proof. The matrix $D = UXU^{-1}$ must be, respectively, Hermitian or unitary. Thus in the respective cases $D = (D^T)^* = D^*$ and $D^{-1} = (D^T)^* = D^*$. □

EXERCISES

Level 1

1. Write these normal matrices as $U^{-1}DU$. To do this, first find the characteristic polynomial. Then find its roots, the eigenvalues. For each eigenvalue k solve the linear equation $vA = vk$ to find an eigenvector $v \neq 0$, (for instance take $v_1 = 1$). Then write U, D as in Corollary 3.7.7. The matrix U need not be unitary.

$$\begin{bmatrix} 1 & 1 \\ 1 & 1 \end{bmatrix}, \begin{bmatrix} 2 & 1 \\ 1 & 2 \end{bmatrix}, \begin{bmatrix} 0 & 2 \\ 2 & 3 \end{bmatrix}, \begin{bmatrix} 1 & 1 \\ -1 & 1 \end{bmatrix}, \begin{bmatrix} 0 & 6 \\ 6 & 5 \end{bmatrix}$$

$$\begin{bmatrix} 1 & 1 & 1 \\ 1 & 1 & 1 \\ 1 & 1 & 1 \end{bmatrix}$$

(Hint. Use as basis $\{(1, 1, 1), (0, 1, -1), (-2, 1, 1)\}$.)

2. Show by calculation that the eigenvectors of this matrix for different eigenvalues are orthogonal.

$$\begin{bmatrix} 0 & 1 & 1 \\ -1 & 0 & 1 \\ -1 & -1 & 0 \end{bmatrix}$$

Level 2

1. Prove directly that if M is symmetric, then two eigenvectors v_1, v_2 belonging to different eigenvalues k_1, k_2 are orthogonal. Show $xMy = k_1 v_1 \cdot v_2 = k_2 v_1 \cdot v_2$. Conclude that $v_1 \cdot v_2 = 0$ since $k_1 \neq k_2$.
2. A quadratic form is an expression of the form

$$\sum_{j=1}^{n} \sum_{i=1}^{n} a_{ij} x_i x_j$$

such as $x_1^2 + 2x_1x_2 - x_2^2$. Prove that the quadratic form can be represented as xAx^{T} where $A = (a_{ij})$, $x = (x_1, x_2, \ldots, x_n)$. Here we assume the a_{ij} are real numbers.
3. Show that in Exercise 2 we can always take the matrix a_{ij} to be symmetric. If $a_{ij} \neq a_{ji}$ replace both by

$$\frac{(a_{ij} + a_{ji})}{2}$$

4. Prove that if we make a substitution of variables $y_1, y_2, \ldots, y_n$ for $x_1, x_2, \ldots, x_n$ according to the rule $x_j = \Sigma\, y_i b_{ij}$ that the quadratic form xAx^{T} goes to $(yB)A(yB)^{\mathrm{T}} = y(BAB^{\mathrm{T}})y^{\mathrm{T}}$.
5. Prove that if A is a real symmetric matrix there exists a real unitary matrix U and a real diagonal matrix D such that $A = U^{-1}DU = U^{\mathrm{T}}DU$ ($U^{\mathrm{T}} = U^{-1}$ by unitariness). By Corollary 3.7.7 each eigenvalue k is real. Thus we can always find real eigenvectors by solving the linear equations $vA = kv$ in which all coefficients are real. (These have a solution v not zero since

$A - kI$ is singular, hence one equation may be deleted.) Now the argument of this section goes through where all numbers are real.

6. Show that any real quadratic form can be expressed in terms of vectors $y_1, y_2, \ldots, y_n$ as $\Sigma\, c_i y_i^2$ for $c_i \in \mathbf{R}$. Here $x_j = \Sigma\, y_i b_{ij}$ and the matrix $B = (b_{ij})$ is orthogonal.
7. Show we can write the quadratic form xy in this form by making the substitution

$$x = \frac{u+v}{\sqrt{2}}, \quad y = \frac{u-v}{\sqrt{2}}$$

8. Write the quadratic form $xy + yz + xz$ as $\Sigma\, c_i w_i^2$ where $c_i \in \mathbf{R}$.

Level 3

1. Let M be a normal matrix. $MM^{T*} = M^{T*}M$ is real symmetric, and its eigenvalues are the squares of the absolute values of the eigenvalues of M.
2. Show a real symmetric matrix M has a real symmetric square root X such that $M = XX$ provided that all eigenvalues of X are nonnegative. Use $M = UDU^{-1}$ and find $\sqrt{D}$.
3. Show any normal matrix A can be written as BC where $BC = CB$, B is Hermitian and C is unitary. Let $A = UDU^{-1}$ and factor $D = XY$ where $x_{ii} = |d_{ii}|$,

$$y_{ii} = \frac{d_{ii}}{x_{ii}}.$$

Let $B = UXU^{-1}$, $C = UYU^{-1}$. This is called the ***polar decomposition*** of a normal matrix.

4. The signature of a quadratic form is the number of positive eigenvalues of its matrix minus the number of negative eigenvalues. Show two quadratic forms of the same signature, where the matrices A are both nonsingular, can be converted one into the other by a substitution of the form BAB^{T}, $b_{ij} \in \mathbf{R}$ and $\det(B) \neq 0$. Use Level 2, Exercise 6 to obtain the form $\Sigma\, c_i y_i^2$ where the c_i are the eigenvalues.
5. Show that the signature of a real quadratic form is invariant under substitution $A \to BAB^{T}$ for B real and nonsingular.
6. A quadratic form xAx^{T} is called ***positive definite*** if $xAx^{T} > 0$ for all $x \neq 0$. Show this is equivalent to all eigenvalues of A being positive.

CHAPTER 4

Rings

A ***ring*** is a general system in which there are two operations, addition and multiplication. A ring is a commutative group under addition, a semigroup under multiplication, and satisfies distributive laws. We describe the integers as a special ring in this chapter. The set of $n \times n$ matrices over a field and the set of polynomials in a variables over a field are also rings.

There are many kinds of rings. An integral domain is a ring having a unit 1, satisfying the commutative law $ab = ba$ and a cancellation property that if $ac = bc$, $c \neq 0$ then $a = b$. A Euclidean domain is an integral domain in which we can divide a nonzero element y into an element x to obtain a quotient and a remainder such that the remainder is of smaller degree than y. The integers are an example of both. In any Euclidean domain prime numbers and divisibility have most of the usual properties. Every element can be uniquely factored into primes.

An ideal in a ring R is an additive subgroup H such that if $x \in R$, $y \in H$ then $xy, yx \in H$. For any ideal there exists a congruence defined by $x \sim y$ if and only if $x - y \in H$. The equivalence classes form the quotient ring R/H. In the case of the integers these concepts take the form that $x \equiv y(\text{mod } m)$ if $x - y$ is a multiple of m. For instance $5 \equiv 1(\text{mod } 2)$ since 2 divides $5 - 1 = 4$. Congruences can be added, subtracted, and multiplied. The quotient ring is a finite ring $\mathbf{Z}_m$ of m elements. For m prime, it is a field.

Two numbers x, y are called ***relatively prime*** if they have no common divisor (c.d.) except 1.

An element $\bar{c}$ in $\mathbf{Z}_m$ has an inverse $(\bar{c})^{-1}$ if and only if c is relatively prime to m. The elements with inverses form a group. Its order is $\phi(m)$, the number of positive integers from 1 to $n - 1$ relatively prime to m. From this follows the Euler-Fermat Theorem $x^{\phi(m)} \equiv 1(\text{mod } m)$ if x is relatively prime to m. For m prime the group of nonzero elements of $\mathbf{Z}_m$ is cyclic. This gives a criterion for equations $x^k \equiv c(\text{mod } m)$ to be solvable.

In Section 5 we present an advanced topic, simple and semisimple rings, needed for the study of group representations. A ring is simple if it has no nonzero two-sided ideals. Under a finite dimensionality assumption every simple ring is isomorphic to a complete ring of $n \times n$ matrices. A semisimple ring is one which is a direct sum of simple rings. Proofs of results about semisimplicity are omitted. These results provide the most important classification theory in ring theory. If two-sided ideals I do exist then a study of ideals I and the quotient rings R/I gives information about R.

4.1 THE INTEGERS AND DIVISIBILITY

In this section we state a typical set of axioms for $\mathbf{Z}$ and then prove elementary facts about divisibility. We view $\mathbf{Z}$ here as one case of a general class of systems.

DEFINITION 4.1.1. A *ring* is a set S on which two binary operations $S \times S \to S$ are defined, denoted as addition and multiplication having element 0 such that for all $a, b, c \in S$:

(1) $(a+b)+c = a+(b+c)$, (2) $(ab)c = a(bc)$

(3) $a(b+c) = ab+ac$, (4) $(b+c)a = ba+ca$

(5) $a+b = b+a$, (6) $a+0 = a$

(7) for all a there exists $-a$ such that $a+(-a) = 0$

For brevity, let R denote an arbitrary ring.

In other words a ring is an abelian group under addition and a semigroup under multiplication. The two operations are linked only by the right and left distributive laws (3) and (4).

EXAMPLE 4.1.1. $\mathbf{Z}$ forms a ring.

EXAMPLE 4.1.2. Any field is a ring.

EXAMPLE 4.1.3. $M_n(F)$ forms a ring.

DEFINITION 4.1.2. A ring R is a *ring with unit* if there exists an element $1 \neq 0$ such that $1x = x1$ for all $x \in R$. It is *commutative* if $ab = ba$ for all $a, b \in R$.

EXAMPLE 4.1.4. A field is a commutative ring with unit.

DEFINITION 4.1.3. A ring R is *ordered* if there exists a strict partial order $<$ on R such that if $x < y$ then (1) for all $z \in R$, $x+z < y+z$, (2) for all $z > 0$ in R, $xz < yz$ and $zy < zx$.

EXAMPLE 4.1.5. **Z**, **Q**, **R** are ordered rings.

A ring is ***linearly ordered*** if the order $<$ is also a linear order, that is, for all $x, y, z \in \mathcal{R}$ either $x = y$, $x < y$, or $x > y$. The symbol $>$ is defined by $x > y$ if and only if $y < x$.

The axioms for **Z** can now be stated compactly.

AXIOM 1. The integers are a linearly ordered, commutative ring with unit.
AXIOM 2. Let $S \subset \mathbf{Z}$ be such that (i) $1 \in S$ and (ii) if $x \in S$ then $x + 1 \in S$. Then S contains all positive integers $\mathbf{Z}^+$ (all x such that $0 < x$).

Axiom 2 is called the ***inductive axiom***. All the usual forms of mathematical induction follow from it.

DEFINITION 4.1.4. In $\mathcal{R}$, $a - b$ is $a + (-b)$ for all $a, b \in \mathcal{R}$.

DEFINITION 4.1.5. The *absolute value* $|x| = x$ if x is positive or zero and $|x| = -x$ if x is negative.

Next we state a number of simple properties of **Z** which follow from the axioms, without proof. (The proofs are in the exercises.) Let $a, b, c \in \mathbf{Z}$.

Z1. $1 > 0$.
Z2. $(-a)(b) = a(-b) = -(ab)$.
Z3. $(-1)(-1) = 1$.
Z4. $a(b - c) = ab - ac$.
Z5. If a, b are positive so is ab.
Z6. If one a, b is positive and the other is negative $ab < 0$.
Z7. If a, b are negative, ab is positive.
Z8. $a0 = 0$.
Z9. If $a > 0$ then $-a < 0$.
Z10. If $a < 0$ then $-a > 0$.
Z11. If $a + b = a + c$ then $b = c$.
Z12. If $ab = 0$ then $a = 0$ or $b = 0$.
Z13. $|a| \geqslant 0$.
Z14. If $a \neq 0$ then $|a| > 0$.
Z15. $|ab| = |a|\,|b|$.
Z16. $|a| + |b| \leqslant |a| + |b|$.
Z17. The smallest positive integer is 1.
Z18. If $a \neq 0$ then $aa > 0$.
Z19. If $ab = 1$ then $a = 1$ or $a = -1$, and $a = b$.
Z20. $|a| = |-a|$.

DEFINITION 4.1.6. An *integral domain* is a commutative ring with unit in which if $ab = 0$ then $a = 0$ or $b = 0$. For brevity, $\mathcal{D}$ will denote an arbitrary integral domain.

EXAMPLE 4.1.6. The integers, any field, and any ***subring*** of a field (subset forming a ring under the same operations) are integral domains.

For the rest of this section, $\mathbf{F}[x]$ ($\mathbf{Q}[x]$) will denote the ring of polynomials over $\mathbf{F}$ ($\mathbf{Q}$).

PROPOSITION 4.1.1. *Let $a, b, c \in \mathcal{D}$. If $ac = bc$, $c \neq 0$, then $a = b$.*

Proof. Since $ac - bc = (a - b)c = 0$, $c = 0$ or $a - b = 0$. So $a - b = 0$. □

DEFINITION 4.1.7. If $a, b \in \mathcal{D}$, *a divides b* if and only if there exists $c \in \mathcal{D}$ such that $a = bc$. This is written as $a | b$.

EXAMPLE 4.1.7. In the ring of polynomials with integer coefficients, $x^2 - 1 | x^4 - 1$.

DEFINITION 4.1.8. If $a \in \mathcal{D}$ and a has as inverse in $\mathcal{D}$, then a is called a *unit*.

EXAMPLE 4.1.8. The units of the integers are precisely ± 1.

EXAMPLE 4.1.9. The units of $\mathbf{F}[x]$ are the nonzero elements of $\mathbf{F}$.

DEFINITION 4.1.9. If $a_1, a_2, \ldots, a_k \in \mathcal{D}$, then $d \in \mathcal{D}$ is called a *greatest common divisor* (*g.c.d.*) of $a_1, a_2, \ldots, a_k$ if and only if (1) $d | a_i$ for $i = 1$ to k, (2) if $x | a_i$ for $i = 1$ to k then $x | d$. The g.c.d. of a and b, is denoted by (a, b).

EXAMPLE 4.1.10. In $\mathbf{Q}[x]$, x is a g.c.d. of $x^2 - x$ and x^3.

PROPOSITION 4.1.2. *The relation $a | b$ is a quasiorder. We have $a | b$ and $b | a$ if and only if $a = ub$ where u is a unit. If $b | a_i$ for $i = 1$ to k then for any $c_1, c_2, \ldots, c_k \in \mathcal{D}$, $b | c_1 a_1 + c_2 a_2 + \ldots + c_k a_k$.*

Proof. Since $a = a1$, $a | a$. If $ra = b$ and $sb = c$ then $rsa = sb = c$. Therefore if $a | b$ and $b | c$, $a | c$.

Let $a | b$ and $b | a$. If a or b is zero both are zero, and $0 = 1(0)$. Suppose a, b are nonzero. Let $a = ub$, $b = va$. Then $a = u(va)$. So by Proposition 4.1.1, $uv = 1$. So u is a unit.

Let $a_i = u_i b$. Then $c_1 a_1 + c_2 a_2 + \ldots + c_k a_k = (c_1 u_1 + c_2 u_2 + \ldots + c_k u_k) b$. □

The g.c.d. also has a kind of associative property.

PROPOSITION 4.1.3. *Let g be a g.c.d. of a finite set S. Let e be a g.c.d. of a finite set T and let f be a g.c.d. of X where $S = T \cup X$. Then g is a g.c.d. of e, f.*

Proof. Since g is a common divisor (c.d.) of S and $T, X \subset S$, it is a c.d. of T, X. Therefore $g|e$ and $g|f$. If d is a c.d. of e, f then $d|e$ and $d|f$ so d is a c.d. of S, T. So d is a c.d. of $S \cup T$ or X. So $d|g$. □

PROPOSITION 4.1.4. *If d, g are g.c.d. of a set S then $d = ug$ where u is a unit.*

Proof. We have $d|g$ and $g|d$. □

EXERCISES

Level 1

1. Prove Property Z11. Add $-a$ to both sides.
2. Prove Property Z8. Note that $a + 0a = 1a + 0a = (1 + 0)a = 1a = a$. Now add $-a$ to both sides.
3. Prove Property Z2. Note that $ab + (-a)b = (a + (-a))b = 0b = 0$ by Exercise 2. Now add $-(ab)$ to both sides. The relation $a(-b) = -ab$ follows by the commutative law.
4. Prove Property Z6 from Exercise 2 and Property Z2.
5. Prove Property Z9 from Definition 4.1.3, adding $(-a)$ to both sides.
6. Prove Property Z10 from Definition 4.1.3, adding $(-a)$ to both sides.

Level 2

1. Prove Properties Z13, Z14 by considering each of three cases, a is positive, negative, or zero.
2. Prove Property Z20 in that way.
3. Prove Property Z3 using $0 = 00 = (1 + (-1))(1 + (-1)) = 1(1 + (-1)) + (-1(1 + (-1))) = 1 + (-1) + (-1)(1) + (-1)(-1) = 0 + (-1) + (-1)(-1) = -1 + (-1)(-1)$. Add 1 to both sides.
4. Prove Property Z7 using Properties Z2, Z3, Z10.
5. Prove Property Z18 using Properties Z5, Z7 and two cases: a positive or negative.
6. Prove Property Z1. Suppose 1 is negative. Then $1(1) = 1 < 0$. This contradicts Property Z18.
7. Prove Property Z4.

Level 3

1. Prove Property Z6 using Properties Z9, Z10, Z2, Z3.
2. Prove Property Z15 by taking five cases: $a = 0$ or $b = 0$, $a > 0$ and $b > 0$, $a > 0$ and $b < 0$, $a < 0$ and $b > 0$, $a < 0$ and $b < 0$.

3. Prove Property Z12 using any from Properties Z1–Z11 and possibly Properties Z13–Z15.
4. Prove Property Z16 by taking five cases as in Exercise 2.
5. Prove Property Z17 from the induction axiom.
6. Prove Property Z19.
7. Show any ring R is a subring of a ring R_1 with unit. Let $R_1 = \mathbf{Z} \oplus R$ and define $(a, b) + (b, d) = (a + c, b + d)$. Define $(a, b)(c, d) = (ac, ad + bc + bd)$. Show R is a subring of R_1, and R_1 is a ring. What is the unit?
8. Which of the listed properties of **Z** hold for all integral domains?

4.2 EUCLIDEAN DOMAINS AND FACTORIZATION

In this section we consider a special type of integral domain called a ***Euclidean domain*** and show that the integers are a special case of this. We observe that in a Euclidean domain any two elements have a g.c.d., and that every element can be factored uniquely into primes.

DEFINITION 4.2.1. An element $p \in \mathcal{D}$ is *prime* if and only if whenever $p = ab$ for $a, b \in \mathcal{D}$, either a or b is a unit and p itself is not a unit. We say that the integers a and b are *relatively prime* (or that a is prime to b) if $(a, b) = 1$.

EXAMPLE 4.2.1. The numbers 2, 3, 5, 7, 11, 13, 17 are prime.

EXAMPLE 4.2.2. The polynomial $x + a$ is prime for any a, where x is an indeterminate.

A Euclidean domain is an integral domain in which we can divide to obtain a quotient and remainder, where the remainder is in a certain sense less than the divisor.

In dealing with the ring of polynomials in a variable (indeterminate) x over a coefficient field **F**, it is more precise to say that x is transcendental over **F**, that is, it does not satisfy any nonzero polynomial equation with coefficients in **F**. Such a polynomial ring can be constructed by taking a subset of the Cartesian product of a countable number of copies of **F**, and defining operations appropriately.

DEFINITION 4.2.2. The integral domain $\mathcal{E}$ is a *Euclidean domain* if and only if for every $c \in \mathcal{E}$, $c \neq 0$, there exists a nonnegative integer $\nu(c)$ such that for all $a, b \in \mathcal{E}$, $a \neq 0$, $b \neq 0$: (i) $\nu(ab) \geqslant \nu(a)$, (ii) there exist $q, r \in \mathcal{E}$ such that $a = qb + r$ and either $r = 0$ or $\nu(r) < \nu(b)$.

We remark that from (i) it follows that $\nu(a) \geqslant \nu(1)$ for all a.

EXAMPLE 4.2.3. Let $\mathcal{D}$ be the integers and let $\nu(a) = |a|$.

EXAMPLE 4.2.4. Let $\mathcal{D}$ be $\mathbf{F}[x]$ and let $\nu(a)$ be the degree of the polynomial a.

EXAMPLE 4.2.5. Let $\mathcal{D}$ be $\{a + b\sqrt{-1}\colon a, b \in \mathbf{Z}\}$ and let $\nu(x) = |x|^2 = a^2 + b^2$.

THEOREM 4.2.1. *If $a, b \in \mathcal{E}$, $a \neq 0$, $b \neq 0$, then there exists a g.c.d. of a, b such that $d = ra + sb$ for some $r, s \in \mathbf{Z}$.*

Proof. Let $K = \{ax + by : x, y \in \mathcal{E}\}$. Let g be a nonzero element of K such that $\nu(g)$ is as small as possible. Write $g = at + sb$. We must first show $g | a$ and $g | b$. Suppose $g | a$ is false. Then $a = qg + r$ where $\nu(r) < \nu(g)$, and $r \neq 0$. Also $r \in K$ since $r = a - qg = a - atq - sbq = a(1 - tq) + b(-sq)$. This contradicts the assumption that $\nu(g)$ was a minimum. So $g | a$. Likewise $g | b$. Suppose $x | a$ and $x | b$. Then $x | ta + sb$. This proves g is a g.c.d. of a, b. □

PROPOSITION 4.2.2. *If $a_1, a_2, \ldots, a_k$ are nonzero elements of $\mathcal{E}$, then there exists a greatest common divisor g of $a_1, a_2, \ldots, a_k$ of the form $g = a_1x_1 + a_2x_2 + \ldots + a_kx_k$ where $x_1, x_2, \ldots, x_k \in \mathcal{D}$.*

Proof. First find a g.c.d. g_1 of a_1, a_2. Then let g_2 be a g.c.d. of g_1, a_3. Find $g_3, g_4, \ldots, g_{k-1}$ in similar fashion, always using Theorem 4.2.1. Then let $g = g_{k-1}$. We will have $g | a_i$ for $i = 1$ to k and $g = a_1x_1 + a_2x_2 + \ldots + a_kx_k$. This implies g is a g.c.d. of $x_1, x_2, \ldots, x_k$. □

EXAMPLE 4.2.6. In $\mathbf{Q}[x]$, 1 is a g.c.d. of x^3 and $x^2 - 1$. We have $1 = xx^3 + (-x^2 - 1)(x^2 - 1)$.

Since we can multiply both sides of $g = a_1x_1 + a_2x_2 + \ldots + a_kx_k$ by any unit, any g.c.d. can be expressed in this form.

PROPOSITION 4.2.3. *Let a, b be nonzero elements of $\mathcal{E}$. Then $\nu(ab) = \nu(b)$ if and only if a is a unit.*

Proof. Let a be a unit. Then $\nu(ab) \geqslant \nu(b)$. And $\nu(b) = \nu(a^{-1}ab) \geqslant \nu(ab)$. So $\nu(ab) = \nu(b)$. Suppose $\nu(ab) = \nu(b)$. Then $b = qab + r$ for some $q, r \in \mathcal{E}$ where $r = 0$ or $\nu(r) < \nu(ab) = \nu(b)$ where $\nu(x)$ denotes the degree of x. Suppose $r \neq 0$. Then $r = b(1 - qa)$. So $\nu(r) \geqslant \nu(b)$. This is false. So $r = 0$. So $b = qab$. So $qa = 1$. So a is invertible. □

PROPOSITION 4.2.4. *Let $a, b, c \in \mathcal{E}$. If $c | ab$ and $(c, a) = 1$ where $(c, a) = 1$ means that c and a are relatively prime then $c | b$.*

Proof. We have $ax + cy = 1$ for some $x, y \in \mathbf{Z}$. So $abx + cby = b$. Since c divides the left-hand side, it divides the right-hand side. □

COROLLARY 4.2.5. *Let $a, b \in E$. If $p|ab$ and p is prime then $p|a$ or $p|b$.*

PROPOSITION 4.2.6. *Let a be a nonzero element of E which is not a unit. Then there exist primes $p_1, p_2, \ldots, p_r$ for some integer $r > 0$ such that $a = p_1 p_2 \ldots p_r$.*

Proof. Let a be an element of minimum degree for which this assertion fails. Suppose $\nu(a) = \nu(1)$. Then since $\nu(a1) = \nu(1)$, a is a unit by Proposition 4.2.3. This is contrary to hypothesis. So $\nu(a) > \nu(1)$. Since a is not prime, $a = bc$ where neither b nor c is a unit. If $\nu(a) = \nu(b)$ then c would be a unit, by Proposition 4.2.3. So $\nu(b) < \nu(a)$. Likewise $\nu(c) < \nu(a)$. But the proposition is true for all elements of degree less than a. So b and c are products of primes. So $a = bc$ is a product of primes. This completes the proof. □

In most cases it is fairly clear that the last result holds. It is less obvious, however, that a factorization into primes is unique except for rearrangement of the primes and multiplication of each prime by a unit.

We first note that it follows by induction from Proposition 4.2.4, that if p is prime and $p|a_1 a_2 \ldots a_k$ then $p|a_i$ for some i. That is, $p|a_1$ or $p|a_2 a_3 \ldots a_k$. If $p|a_2 a_3 \ldots a_k$, then $p|a_2$ or $p|a_3 a_4 \ldots a_k$. And so on.

THEOREM 4.2.7. *Let a be a nonzero element of E which is not a unit. Let $a = p_1 p_2 \ldots p_m = q_1 q_2 \ldots q_n$ be factorizations of a into primes. Then $m = n$, and we can renumber $q_1 q_2 \ldots q_m$ in such a way that $u_i p_i = q_i$ where u_i is a unit, for $i = 1$ to n.*

Proof. Let k be the minimum of n, m. If $k = 1$ then a is prime. So $m = 1$ and $p_1 = q_1$. Now suppose the theorem is true for $k = 1, 2, \ldots, r$. Let $a = p_1 p_2 \ldots p_{r+1} = q_1 q_2 \ldots q_n$. Then $p_1 | q_1 q_2 \ldots q_n$. So $p_1 | q_i$ for some i. So $p_1 u_1 = q_i$ for some u_1. Since q_i is prime, u_1 must be a unit. Renumber $q_1, q_2, \ldots, q_n$ so that q_i is q_1. Then $p_1 u_1 = q_1$. So also $p_2 p_3 \ldots p_r = (u_1^{-1} q_2) q_3 \ldots q_n$. Since the theorem is true for $k < r + 1$, then $r = n - 1$ and we can renumber $q_2, q_3, \ldots, q_n$ and find $v_1, u_3, \ldots, u_n$ such that $v_2 p_2 = u_1^{-1} q_2$, $u_3 p_3 = q_3, \ldots, u_n p_n = q_n$. Let $u_2 = u_1 v_2$. Then the theorem has been verified. □

This theorem also holds for certain non-Euclidean rings. It is known that polynomials in several variables do not form a Euclidean ring, yet unique factorization still is true.

In any Euclidean domain, there is an effective method: (i) to find the g.c.d. g of two nonzero elements a, b and (ii) to express g as $ax + by$. Label a, b so that $\nu(a) \geqslant \nu(b)$. Let $x_1 = a$, $x_2 = b$. Obtain x_{i+1} by setting $x_{i-1} = q_i x_i + x_{i+1}$. Then $\nu(x_{i+1}) < \nu(x_i)$ unless $x_{i+1} = 0$. So eventually some x_i is zero. The last nonzero element is taken as the g.c.d. g. This procedure is called the *Euclidean Algorithm*.

PROPOSITION 4.2.8. *In any* $\mathcal{E}$ *the last nonzero element g is the g.c.d. of a, b.*

Proof. Let $g = x_i$. Then $x_{i-1} = q_i x_i$. We prove by a backwards induction that for each j, $g|x_j$, $g|x_{j-1}$, and g is a linear combination of x_j, x_{j-1}. For $j = i$ this is immediate. Assume $g|x_j$, $g|x_{j-1}$, and $g = rx_j + sx_{j-1}$. We have

$$x_{j-2} = q_{i-1}x_{j-1} + x_j$$

Therefore g divides x_{j-2} and x_{j-1}. And $g = rx_j + sx_{j-1} = r(x_{j-2} - q_{i-1}x_{j-1}) + sx_{j-1}$. This completes the induction. Thus $g|a$, $g|b$ and $g = ax + by$ for some x, y. Thus g is the g.c.d. □

EXAMPLE 4.2.7. Find (31, 47). Find x and y such that $(31, 47) = 31x + 47y$. Divide and take remainders.

$$47 = 1 \cdot 31 + 16$$

$$31 = 1 \cdot 16 + 15$$

$$16 = 1 \cdot 15 + 1$$

$$15 = 15 \cdot 1 + 0$$

The last nonzero remainder, 1 is the g.c.d. To find the linear combination, solve the equations for the remainders

$$47 - 1 \cdot 31 = 16$$

$$31 - 1 \cdot 16 = 15$$

$$16 - 1 \cdot 15 = 1$$

Now start at the last equation and substitute.

$$1 = 16 - 1 \cdot 15 = 16 - (31 - 16) = 2 \cdot 16 - 31$$

$$= 2(47 - 31) - 31 = 2 \cdot 47 - 3 \cdot 31$$

Simplify but do not alter the numbers in the series x_i.

EXAMPLE 4.2.8. Find the g.c.d. of $x^3 - 1$, $2x^2 - 3x + 1$.

Divide polynomials:

$$\begin{array}{r}
\frac{1}{2}x + \frac{3}{4} \\
2x^2 - 3x + 1 \overline{\smash{)}\, x^3 + \; 0 \; + \; 0 \; - 1} \\
x^3 - \frac{3}{2}x^2 + \frac{1}{2}x \quad \\
\hline
\frac{3}{2}x^2 + \frac{1}{2}x - 1 \\
\frac{3}{2}x^2 - \frac{9}{4}x + \frac{3}{4} \\
\hline
\frac{7}{4}x - \frac{7}{4}
\end{array}$$

We may write $\frac{7}{4}(x-1)$ and use $(x-1)$ as a divisor in the next stage since $\frac{7}{4}$ is a unit when we are dealing with polynomials

$$\begin{array}{r|l} & 2x-1 \\ \hline x-1 & 2x^2-3x+1 \\ & 2x^2-2x \\ \hline & -x+1 \\ & -x+1 \\ \hline & 0 \end{array}$$

The g.c.d. is $(x-1)$.

EXERCISES

Level 1

1. Factor 192 as a product of primes.
2. Find (36, 54).
3. Find the g.c.d. of 501, 111 by the Euclidean Algorithm.
4. Find integers x, y such that $4x + 7y = 1$.
5. Find the g.c.d. of 700, 133.
6. Find the g.c.d. of $x^3 - 4x + 1$ and its derivative.

Level 2

1. Find numbers x, y such that $17x + 11y = 1$.
2. Find numbers x, y, z such that $6x + 15y + 10z = 1$. (First find numbers such that $6r + 15s = 3$. Then find numbers such that $3x + 10z = 1$.)
3. Find polynomials $f(x), g(x)$ such that $f(x)(x^2+1)+g(x)(x^2+x+1) = 1$.
4. If $ar + bs = 1$ (so a, b are relatively prime) show that $a(r + mb) + b(s - ma) = 1$.
5. Show that Exercise 4 yields all solutions of $ax + by = 1$.

Level 3

1. Prove from the properties given in Section 4.1 that $\mathbf{Z}$ is a Euclidean domain, i.e. prove that for any $x \neq 0$, $y \neq 0$ there exist q, r with $x = qy + r$ and $|r| < |x|$. Take r to be a number $x - qy$ of minimum absolute value.
2. Prove that the set of integers $\{a + bi: a, b \in \mathbf{Z}\}$ is a Euclidean domain.
3. Factor $1 + 3i$ into primes in this ring. (Try factoring $a^2 + b^2 = |z|^2$.)
4. Show unique factorization into primes fails in the subring of $\mathbf{Q}[x]$ generated by $1, (x-1)^2, (x-1)^3$.
5. Although unique factorization into primes holds in the ring of polynomials in 2 variables over $\mathbf{C}$, show that this is not a Euclidean domain. To do this show that a g.c.d. of x, y is 1, yet there do not exist $f(x)$, $g(x)$ such that $f(x)x + g(x)y = 1$.

4.3 IDEALS AND CONGRUENCES

An ***ideal*** is a subset I of a ring R which is a subring, and is such that if $a \in I$, $b \in R$ then $ab, ba \in R$. Ideals are central in the study of the structure of rings. For any ideal, the relation $x = y + m, m \in I$ is called a ***congruence.*** A congruence is an equivalence relation for which the equivalence classes themselves form a ring. This ring is frequently simpler, but gives properties of the original ring. It is called a ***quotient ring.*** All nontrivial quotient rings of $\mathbf{Z}$ are finite. Additively they are the $\mathbf{Z_m}$ mentioned in connection with groups.

DEFINITION 4.3.1. In R, a nonempty set I is a *left* (*right, two-sided*) *ideal* if for all $a, b \in I, c \in R$ we have $-a \in I$, $a + b \in I$ and $ca \in I$ ($ac \in I$, $ca, ac \in I$).

EXAMPLE 4.3.1. Let $m \in \mathbf{Z}^+$. Then the set of multiples km of m is an ideal.

EXAMPLE 4.3.2. For any ring R and finite set of elements $a_1, a_2, \ldots, a_n \in R$ the set $Ra_1 + Ra_2 + \ldots + Ra_n = \{r_1 a_1 + r_2 a_2 + \ldots + r_n a_n : r_i \in R\}$ is a left ideal called the ***left ideal generated*** by $a_1, a_2, \ldots, a_n$.

A two-sided ideal is frequently just called an ***ideal.*** For commutative rings there is no difference among left, right, and two-sided ideals.

PROPOSITION 4.3.1. *For any two ideals I, J both either right, left, or two-sided, the following are ideals of the same type:*

(*1*) $I + J = \{x + y : x \in I, y \in J\}$
(*2*) $IJ = \{x_1 y_1 + x_2 y_2 + \ldots + x_n y_n : n \in \mathbf{Z}, x_i \in I, y_i \in J\}$
(*3*) $I \cap J$

For any set S the intersection of all ideals containing S is an ideal (*the ideal generated by S*).

Proof. We verify (1), (2), (3) for left ideals. First the additive property (1) $a_1 + b_1 + a_2 + b_2 = (a_1 + a_2) + (b_1 + b_2)$, (2) $(x_1 y_1 + x_2 y_2 + \ldots + x_n y_n) + (r_1 s_1 + r_2 s_2 + \ldots + r_k s_k)$ is again a sum of this type, (3) $a + b \in I$ and $a + b \in J$ so $a + b \in I \cap J$. For the multiplicative property (1) $c(x + y) = cx + cy \in I + J$, (2) $c(x_1 y_1 + x_2 y_2 + \ldots + x_n y_n) = (cx_1)y_1 + (cx_2)y_2 + \ldots + (cx_n)y_n$, (3) $ca \in I$ and $ca \in J$ so $ca \in I \cap J$.

The remainder of the proof including the verification that $-a$ is in the ideal, is similar. □

In every ring R there are two ideals, $\{0\}$ and R. Since $a + (-a) = 0$, 0 is in every ideal.

DEFINITION 4.3.2. A *congruence* on a ring R is an equivalence relation $x \sim y$ such that if $x \sim y$ then for all z, $x + z \sim y + z$, $xz \sim yz$, and $zx \sim zy$.

All congruences of rings arise in the way given in the next result.

If $x \sim y$ and $z \sim w$ then $x + z \sim x + w \sim y + w$ and $xz \sim xw \sim yw$.

PROPOSITION 4.3.2. *The relation* $x \sim y$ *if and only if* $x - y \in \mathcal{I}$ *is a congruence.*

Proof. If $a \sim b$, $b \sim c$, then $a - b \in \mathcal{I}$, $b - c \in \mathcal{I}$. Since $\mathcal{I} + \mathcal{I} \subset \mathcal{I}$, $(a - b) + (b - c) = a - c \in \mathcal{I}$. So $a \sim c$. Also $b - a = -(a - b) \in \mathcal{I}$. And $a - a = 0 \in \mathcal{I}$. So it is an equivalence relation. The congruence properties hold also. □

The equivalence classes under any congruence form a ring called a ***quotient ring***, and denoted in the present case by $\mathcal{R}/\mathcal{I}$.

The function $f(x) = \bar{x}$ which assigns to each element its equivalence class is a homomorphism of rings, that is $f(x + y) = f(x) + f(y)$ and $f(xy) = f(x)f(y)$.

For the integers these ideas take the following form.

DEFINITION 4.3.3. For a, $b \in \mathbf{Z}$ we have $a \equiv b(\text{mod } m)$ if and only if $m | a - b$. Here $m | a - b$ means m divides $a - b$. That is, $a - b$ belongs to the ideal $\{km\}$, $k \in \mathbf{Z}$.

EXAMPLE 4.3.3. $7 \equiv 1(\text{mod } 3)$ since $3 | 7 - 1$.

The relation of congruence is an equivalence relation, and it follows from the preceding theory that if $x \equiv y(\text{mod } m)$, $z \equiv w(\text{mod } m)$ we have $x + z \equiv y + w(\text{mod } m)$, $xz \equiv yw(\text{mod } m)$.

The quotient ring is denoted $\mathbf{Z_m}$.

PROPOSITION 4.3.3. *The classes* $\bar{0}, \bar{1}, \bar{2}, \ldots, \overline{m - 1}$ *are distinct and include all congruence classes modulo* m.

Proof. For $x \in \mathbf{Z}$, $x = qm + r$ where $0 \leqslant r < m - 1$. Then $x - r = qm$ so $x \equiv r(\text{mod } m)$. So any x lies in one of these classes. Suppose $0 \leqslant r < s < m - 1$. Then $m > s - r > 0$. Therefore m cannot divide $s - r$. So s, r must lie in distinct classes. □

COROLLARY 4.3.4. $\mathbf{Z_m}$ has exactly m elements.

EXAMPLE 3.3.4. The addition and multiplication tables of $\mathbf{Z_m}$ can be found by taking ab or $a + b$ and then finding the remainder when this is divided by m. For example modulo 5, $\bar{2} + \bar{3} = \bar{5} = \bar{0}$ since

$$\begin{array}{r} 5 \\ 5\,\overline{)\,5} \\ \underline{5} \\ 0 \end{array}$$

and $\overline{2}\,\overline{3} = \overline{6} = \overline{1}$ since

$$\begin{array}{r} 1 \\ 5\overline{)6} \\ \underline{5} \\ 1 \end{array}$$

An ideal is called *proper* if it is a proper subset of the ring. It is called *trivial* if it equals $\{0\}$.

The proof of the following result follows the same pattern as for groups (it is not given in full).

THEOREM 4.3.5. *Let $f: R_1 \to R_2$ be a homomorphism from a ring R_1 onto a ring R_2 with kernel $K = \{x : f(x) = 0\}$. Then K is an ideal and f gives an isomorphism $R_1/K \to R_2$.*

Proof. If $x - y \to K$ then $f(x) - f(y) = 0$. So f gives a well-defined map $\overline{f}$ on equivalence classes from $R_1/K \to R_2$. This is a ring homomorphism: $\overline{f}(\overline{x}\,\overline{y}) = f(xy) = f(x)f(y) = \overline{f}(\overline{x})\overline{f}(\overline{y})$. It is onto since f is. It is also 1-1. □

EXERCISES

Level 1

1. Work out the mod 5 addition table.
2. Work out the mod 5 multiplication table.
3. Work out the addition and multiplication mod 7.
4. Work out the tables for $\mathbf{Z}_{13}$.
5. Find x such that $4x \equiv 2 (\text{mod } 7)$.

Level 2

1. For any congruence $x \sim y$ on a ring show the set $\{x : x \sim 0\}$ is an ideal.
2. Show the kernel $\{x : f(x) = 0\}$ of any ring homomorphism is an ideal.
3. Show that a field $\mathbf{F}$ has no ideals except $\{0\}$ and $\mathbf{F}$.
4. Prove the addition table of $\mathbf{Z}_m$ has exactly one copy of each element in each row and column. Excluding the row and column of zero, when will this hold for multiplication?
5. Define a ring structure on a Cartesian product $R_1 \times R_2$ of two rings. Show that if both have units there must exist ideals of $R_1 \times R_2$ other than $\{0\}, R_1 \times R_2$.
6. Prove that the last statement of Proposition 4.3.1 agrees with the previous idea of left ideal generated by a set.

Level 3

1. Show any congruence on $\mathcal{R}$ is determined by a unique ideal.
2. Prove any ideal in a Euclidean domain $\mathcal{E}$ is of the form $\mathcal{R}x$ for some element $x \in \mathcal{R}$. Such ideals are called ***principal***.
3. Prove that the family of ideals of $\mathcal{R}$ is a lattice under inclusion.
4. Prove that any integral domain $\mathcal{D}$ is a subring of a **F**. Define **F** by taking ordered pairs (a, b) from **R** interpreted as fractions a/b. On these ordered pairs take a congruence $(a, b) \sim (c, d)$ if $ad = bc$ and use the usual definition for sums and products of fractions.
5. Let $\mathcal{I} \neq \{0\}, \mathcal{I} \neq \mathcal{R}$, and $\mathcal{I}$ is maximal under inclusions among proper ideals of $\mathcal{R}$. Prove $\mathcal{R}/\mathcal{I}$ is a field.
6. Prove that if $\mathcal{R}/\mathcal{I}$ is a field, then $\mathcal{I}$ cannot be contained in another proper ideal $\mathcal{J}$.

4.4 STRUCTURE OF $\mathbf{Z_n}$

The rings $\mathbf{Z_n}$ have a number of special properties. They are fields for n prime. Algebraic equations can be considered in them, such as $ax = b$ or $x^2 = c$. Various identities hold, of which the most famous is Fermat's Theorem.

PROPOSITION 4.4.1. *In* $\mathbf{Z_m}$, *an element* $\bar{c}$ *has a multiplicative inverse if and only if* $(c, m) = 1$.

Proof. If $\bar{c}x = \bar{1}$ then $cx - 1 = km$, $cx - km = 1$. So c, m cannot have a common divisor. Suppose c, m are relatively prime. Then $cx - km = 1$ for some x, m. Therefore $cx \equiv 1 (\text{mod } m)$. □

COROLLARY 4.4.2. $\mathbf{Z_m}$ *is a field if and only if m is prime.*

In any ring the set of elements having inverses is a group since we can take $(x^{-1})^{-1} = x$ and $(xy)^{-1} = y^{-1}x^{-1}$. Thus the set of invertible elements of $\mathbf{Z_m}$ forms a group.

DEFINITION 4.4.1. The *Euler function* $\phi(m)$ is the number of integers $x, 0 < x < m$ which are relatively prime to m.

EXAMPLE 4.4.1. If m is p^n, a power of a prime, $\phi(m)$ consists of all numbers $1, 2, \ldots, p^n$ except the p^{n-1} multiples of p. Therefore $\phi(p^n) = p^n - p^{n-1}$.

PROPOSITION 4.4.3. (Euler-Fermat.) *If x and m are relatively prime then* $x^{\phi(m)} \equiv 1 (\text{mod } m)$.

Proof. The order of any element of a group divides the order of the group. □

COROLLARY 4.4.4. *If p is prime then $x^p \equiv x (mod\ p)$ for all x.*

There exist finite fields other than $\mathbf{Z_p}$ as will be shown later.

THEOREM 4.4.5. *Every finite subgroup of the multiplicative group of* **F** *is cyclic.*

Proof. Let G be a multiplicative subgroup of a field of order n. If G has elements a of order k then k divides n. Let a have maximal order k. Suppose b has order t and t does not divide k. Then some prime p occurs to a higher power in t than in k. Let $t = p^s u$, $k = p^w v$ where p does not divide u and p does not divide v. Then b^u has order p^s and a^{p^w} has order v. And $b^u a^{p^w}$ has order $p^s v > k = p^w v$. (Its order divides $p^s v$ but not $p^s v$ divided by q for any prime q.) This contradicts maximality of the order of a.

Therefore the order of b divides k. If $k = n$ then a generates a cyclic subgroup of order n which must coincide with G. Suppose $k < n$. Then there are $n > k$ roots $r_1, r_2, \ldots, r_n$ of the equation $x^k = 1$. But this means $(x - r_1)$, $(x - r_2), \ldots, (x - r_n)$ divide $x^k - 1$. Therefore by unique factorization in $\mathbf{F}[x]$, $(x - r_1)(x - r_2) \ldots (x - r_n)$ divides $(x^k - 1)$. But a polynomial of degree n cannot divide a polynomial of degree $k < n$. □

COROLLARY 4.4.6. *For any prime p there exists an element a such that $a, a^2, \ldots, a^{p-1}$ ranges over all nonzero elements of* $\mathbf{Z_p}$.

DEFINITION 4.4.2. A number x is said to be a *quadratic residue modulo m* if and only if $y^2 \equiv x (\text{mod } m)$ for some $y \in \mathbf{Z_m}$.

EXAMPLE 4.4.2 The number 4 is a quadratic residue to any modulus since $4 \equiv 2^2$.

PROPOSITION 4.4.7. *For p prime and odd, x is a quadratic residue of p if and only if*

$$x^{\frac{p-1}{2}} = 1.$$

Proof. If $x = b^2$ then

$$x^{\frac{p-1}{2}} \equiv b^{p-1} \equiv 1 (\text{mod } p)$$

Let $x = a^k$ for a generator of k of the multiplicative group. If

$$x^{\frac{p-1}{2}} = 1$$

then $(p-1)$ divides $(k \frac{p-1}{2})$ so k is even. So

$$x \equiv (a^{\frac{k}{2}})^2$$

□

COROLLARY 4.4.8. *For p prime, -1 is a quadratic residue of p if and only if $p \equiv 1 (mod\ 4)$.*

DEFINITION 4.4.3. An element $\omega \in \mathbf{F}$ is a *primitive nth root of unity* if $\omega^n = 1$ but $\omega^i \neq 1$ for $i = 1, 2, \ldots, n-1$.

EXAMPLE 4.4.3. The numbers $\pm i$ are primitive 4th roots of unity.

EXERCISES

Level 1

1. A test for primeness of a number p is to check whether $x^{p-1} \equiv 1 (\text{mod } p)$. (It is not always sufficient but for p large gives strong evidence.) Try this for $x = 2, p = 1$ to 10.
2. When does $ax \equiv b (\text{mod } m)$ have a solution x?
3. Find a multiplicative inverse of 5, mod 13.
4. If x, y are quadratic residues modulo m, prove xy is also a quadratic residue modulo m.
5. How many quadratic residues exist modulo 3? 5? Try to generalize to any p.
6. Do you think a product of two quadratic nonresidues must be a quadratic residue, modulo a prime? Compute a number of examples.
7. Prove $x^2 \equiv 1 (\text{mod } 8)$ for any odd x by checking all cases.

Level 2

1. In any field, show by Theorem 4.4.5 that ± 1 are the only solutions of $x^2 = 1$.
2. Show from Exercise 1 that for any x, $x^{\frac{p-1}{2}} \equiv \pm 1$.
3. Prove Wilson's Theorem that $(p-1)! \equiv (-1)^p (\text{mod } p)$ for any prime p. All numbers $1, 2, \ldots, p-1$ are solutions of $x^{p-1} - 1$. Hence $(x-1)(x-2) \ldots (x-p+1)$ divides $x^{p-1} - 1$. Hence $x^{p-1} - 1 = k(x-1)(x-2) \ldots (x-p+1)$. By considering the x^{p-1} term, $k = 1$.
4. Show that if p divides $x^2 + y^2$ and p does not divide x, y then $p \equiv 1 (\text{mod } 4)$, using Corollary 4.4.8.
5. When does the equation $x^2 + ax + b \equiv 0 (\text{mod } p)$ have a root x, for odd primes p?
6. Prove $x^2 + y^2 + z^2$ cannot be congruent to $7 (\text{mod } 8)$, by considering cases.
7. Let the ***Legendre symbol*** $(x \mid y)$ be ± 1 accordingly, as x is or is not a quadratic residue of y. Prove

$$(x \mid y) \equiv x^{\frac{y-1}{2}} (\text{mod } y)$$

for y prime.

8. For primes x, y, there is a relation between $(x|y)$ and $(y|x)$. Experiment to find this (but don't try to prove it). It is ***Gauss's Law of Quadratic Reciprocity.***

Level 3

1. Show that if $f(x)$ and $g(x)$ are polynomials of degree $<p$ in a variable x over $\mathbf{Z_p}$ and $f(k) = g(k)$ for $k = 0, 1, 2, \ldots, p-1$ then $f(x) = g(x)$. Since $k^p \equiv k$, this is false for degree p.
2. Prove for m, r relatively prime $\mathbf{Z_{mr}} \simeq \mathbf{Z_m} \oplus \mathbf{Z_r}$. Map $\mathbf{Z_{mr}} \to \mathbf{Z_m}$ and $\mathbf{Z_{mr}} \to \mathbf{Z_r}$ by the map sending $\bar{k}$ to $\bar{k}$. This is a homomorphism because $\mathbf{Z_{mr}} \subset \mathbf{Z_m}$ and $\mathbf{Z_{mr}} \subset \mathbf{Z_r}$. Show the kernel of this map is zero (this proves isomorphism).
3. Extend Exercise 2 to k factors.
4. Prove the '***Chinese Remainder Theorem***'. If $x_i \equiv c_i \pmod{m_i}$ are a set of congruences, they have a simultaneous solution if and only if for all i, j, $c_i \equiv c_j \pmod{d}$ where $d = \text{g.c.d.}(m_i, m_j)$.
5. Prove

$$\phi(p_1{}^{n_1} p_2{}^{n_2} \ldots p_k{}^{n_k}) = p_1{}^{n_1} p_2{}^{n_2} \ldots p_k{}^{n_k} (1 - \frac{1}{p_1})(1 - \frac{1}{p_2}) \ldots \ldots (1 - \frac{1}{p_k})$$

6. Discuss the multiplicative structure of $\mathbf{Z_{p^n}}$ where p is a prime number and $n \in \mathbf{Z}^+$.
7. Find in a book and write out in your own words a proof of the law of quadratic reciprocity.
8. Prove any finite integral domain is a field.

4.5 SIMPLE AND SEMISIMPLE RINGS

A ***simple ring*** is one with no ideals other than itself and zero. We deal with simple rings which are algebras over the complex numbers of finite dimension. Such algebras it will be shown, are always precisely the rings of $n \times n$ matrices over $\mathbf{C}$, for some n.

DEFINITION 4.5.1. A ring R with unit is an ***F-algebra*** for $\mathbf{F}$ if R is a vector space over $\mathbf{F}$ and for all $a, b \in R$, $\alpha \in \mathbf{F}$ we have $(\alpha a)b = \alpha(ab) = a(\alpha b)$.

EXAMPLE 4.5.1. If $\mathbf{F_2}$ is a field and $\mathbf{F_1}$ is a field containing $\mathbf{F_2}$ then $\mathbf{F_1}$ is an F-algebra. Thus the complex numbers are an F-algebra over $\mathbf{R}$.

EXAMPLE 4.5.2. The ring of $M_n(F)$ is an F-algebra, where $a \in \mathbf{F}$ is sent to aI.

An F-algebra is necessarily an F-vector space. It is called ***finite dimensional*** if it is finite dimensional as a vector space over $\mathbf{F}$.

DEFINITION 4.5.2. A ring R with unit is a *division ring*, if and only if every nonzero element has a multiplicative inverse, and is sometimes denoted $R_{\mathcal{D}}$.

EXAMPLE 4.5.3. A commutative division ring is precisely a field.

EXAMPLE 4.5.4. The quaternions, an F-algebra with basis $\{1, \boldsymbol{i}, \boldsymbol{j}, \boldsymbol{k}\}$ where $\boldsymbol{i}^2 = \boldsymbol{j}^2 = \boldsymbol{k}^2 = -1$, $\boldsymbol{ij} = \boldsymbol{k}$, $\boldsymbol{ji} = -\boldsymbol{k}$, $\boldsymbol{jk} = \boldsymbol{i}$, $\boldsymbol{kj} = -\boldsymbol{i}$, $\boldsymbol{ki} = \boldsymbol{j}$, $\boldsymbol{ik} = -\boldsymbol{j}$, are a division ring for any field $\mathbf{F} \subset \mathbf{R}$.

Over general fields there exist many complicated division algebras. However, we will show next that over $\mathbf{C}$, the only finite dimensional $R_{\mathcal{D}}$ is $\mathbf{C}$ itself.

THEOREM 4.5.1. *Let $R_{\mathcal{D}}$ as a finite dimensional division ring which is a C-algebra. Then* $R_{\mathcal{D}} = \mathbf{C}$.

Proof. Suppose $x \in R_{\mathcal{D}} \backslash \mathbf{C}$. The powers $1, x, x^2, \ldots, x^n, \ldots$ must be linearly dependent, since $R_{\mathcal{D}}$ is finite dimensional. In particular let x^k be the least power of x which is a $\mathbf{C}$-linear combination of $1, x, x^2, \ldots, x^k$. Then $x^k = c_0 + c_1 x + c_2 x^2 + \ldots + c_k x^{k-1}$. So $x^k - c_k x^{k-1} - \ldots - c_0 = f(x) = 0$. And $f(x)$ is the polynomial of least degree satisfied by x. But over $\mathbf{C}$, $f(x)$ factors into linear factors $(x - r_i)$, where r_i are the complex roots of $f(x)$. Thus $(x - r_i) \neq 0$ for all i since $x \notin \mathbf{C}$ but the product of $x - r_i$ is zero. This contradicts the fact $R_{\mathcal{D}}$ is a division ring since if $x - r_i$ have inverses, their product has an inverse and so is nonzero. □

DEFINITION 4.5.3. An element x of R with unit is an *idempotent* if $x^2 = x$. It is a *central idempotent* if also $xy = yx$ for all $y \in R$.

EXAMPLE 4.5.5. In any ring with unit, 1 is a central idempotent.

EXAMPLE 4.5.6. Any diagonal $(0, 1)$-matrix is an idempotent.

DEFINITION 4.5.4. R is ***regular*** if it is regular as a semigroup. That is for all $x \in R$ there exists $y \in R$ such that $xyx = x$.

EXAMPLE 4.5.7. $R_{\mathcal{D}}$ is regular.

EXAMPLE 4.5.8. $M_n(F)$ is regular.

A proof of the following result, which is basically the first Wedderburn theorem, is too technical for this book.

THEOREM 4.5.2. *For a finite dimensional F-algebra A the following are equivalent:*

(1) *A is a direct sum of simple F-algebras.*
(2) *A as a vector space, is a direct sum of left ideals (or right ideals).*
(3) *Every left and right ideal of A is generated by an idempotent.*
(4) *A has no nonzero nilpotent two-sided ideals.*
(5) *A has no nonzero nilpoint left or right ideals.*
(6) *The multiplicative semigroup of A is a regular semigroup.*

An algebra satisfying any of these conditions is called ***semisimple***.

EXAMPLE 4.5.9. Any division algebra is regular, as is the ring of $M_n(F)$.

EXAMPLE 4.5.10. The ring of lower triangular matrices over **F** has a nilpotent ideal (those with zeros on the main diagonal) and so it is not regular.

The proof of the following theorem follows M. Hall (1959).

THEOREM 4.5.3. (Wedderburn.) *A simple finite dimensional algebra over* **F** *is isomorphic to the ring $M_n(A_D)$ of $n \times n$ matrices over a division algebra A_D over* **F** *for some n, A_D.*

Proof. We first find an expression of R as a direct sum of minimal ideals I_k. Let I_1 be a minimal right ideal. Let e_{11} be an idempotent such that $e_{11}I_1 = I_1$. Suppose we have obtained idempotents $e_{11}, e_{22}, \ldots, e_{kk}$ such that (1) $e_{ii}R$ is a minimal right ideal I_i in R, (2) $e_{ii}e_{jj} = 0$ for $i \neq j$. It follows from (1), (2) that $I_j \cap (\sum_{i \neq j} I_i) = 0$ since if x is in the intersection $e_{jj}x = x$ since $x \in I_j$ but $e_{jj}x = 0$ by (2). Suppose $R \neq \Sigma I_i$. Let $x \in R \backslash \Sigma I_i$. Then

$$(1 - \sum_{i=1}^{k} e_{ii})x \neq 0$$

So

$$(1 - \sum_{i=1}^{k} e_{ii})R \neq 0.$$

Let I_{k+1} be a minimal right ideal in this ring. Write $I_{k+1} = wI_{k+1}$ where w is an idempotent in I_{k+1}. Thus $e_{ii}w = 0$ for $i = 1$ to k. Let

$$x = w(1 - \sum_{i=1}^{k} e_{ii})$$

Then $x^2 = w \cdot 1 \cdot x = x$ and $e_{ii}x = 0$ and $xe_{ii} = 0$, So let $e_{k+1,k+1} = x$. This continues an induction. So we can write R as the direct sum of $e_{ii}R$ where e_{ii} is an idempotent and $e_{ii}e_{jj} = 0$ for $i \neq j$. From the direct sum expression we have that $\Sigma e_{ii}y = y$ for all y in R. Thus Σe_{ii} is a left identity. So it will be the two-sided identity 1 of R (take $y = 1$).

The rings $e_{ii}Re_{ii}$ will be the required division rings. We show that $e_{ii}Re_{ii}$ is a division ring. It is a subring with identity e_{ii}. Suppose it is not a division ring. If $y \neq 0$ is not invertible in it then $ye_{ii}Re_{ii} \neq e_{ii}Re_{ii}$. Therefore $ye_{ii}R \neq e_{ii}R$. This contradicts the fact that $e_{ii}R$ was a minimal right ideal. Therefore $e_{ii}Re_{ii}$ is a division ring $R_{\mathcal{D}_i}$.

Next we observe that $e_{jj}Re_{ii} \neq 0$. Since R is simple, $R = Re_{ii}R$. Thus $e_{jj} = \Sigma\, a_k e_{ii} b_k$. So $\Sigma\, e_{jj} a_k e_{ii} b_k \neq 0$. This implies $e_{jj} a_k e_{ii} \neq 0$ for some k.

Choose for $i = 2$ to n a nonzero element $e_{11} b_i e_{ii}$ and write it as e_{1i}. We have $e_{11}e_{1i} = e_{1i}e_{ii} = e_{1i}$. Also $e_{1i}R \subset e_{11}R$ so $e_{1i}R = e_{11}R$ by minimality. There exists therefore y_i such that $e_{1i}y_i = e_{11}$. Set $e_{i1} = e_{ii}y_ie_{11}$. Then $e_{ii}e_{i1} = e_{i1}e_{11} = e_{i1}$, $e_{1i}e_{i1} = e_{1i}(e_{ii}y_ie_{11}) = e_{1i}y_ie_{11} = e_{11}$. Therefore $(e_{1i}e_{i1})^2 = e_{11}$, so $e_{i1}e_{1i} \neq 0$. Moreover $(e_{i1}e_{1i})^2 = e_{i1}e_{1i}$ is a nonzero idempotent in $R_{\mathcal{D}_i}$. So $e_{i1}e_{1i} = e_{ii}$.

Now set $e_{ij} = e_{i1}e_{1j}$, which is consistent with the previous definitions. We have $e_{ij}e_{jk} = e_{i1}e_{1j}e_{j1}e_{1k} = e_{i1}e_{11}e_{1k} = e_{i1}e_{1k} = e_{ik}$. And for $j \neq r$, $e_{ij}e_{rs} = e_{ij}e_{jj}e_{rr}e_{rs} = 0$. Therefore e_{ij} have precisely the properties required to be the 0, 1)-matrices whose only 1 entry is in location i, j.

All the division rings $R_{\mathcal{D}_i}$ are isomorphic to $R_{\mathcal{D}_1}$ by the mapping $y \to e_{1i}ye_{i1}$. And in fact $e_{ii}Re_{jj}$ is isomorphic as a vector space to $R_{\mathcal{D}_1}$ under the mapping $x \to e_{1i}xe_{j1}$ which has an inverse $x \to e_{i1}xe_{1j}$.

It readily follows that we have a homomorphism from the ring of matrices $D = (d_{ij})$ over $R_{\mathcal{D}_1}$ into R sending (d_{ij}) to $\Sigma\, e_{i1}d_{ij}e_{1j}$.

This has kernel zero since if we multiply by e_{ii} and e_{jj} on left and right only the term in d_{ij} will remain. It is onto since $x = 1(x)1 = (\Sigma\, e_{ii})(x)(\Sigma\, e_{ii})$ and $e_{ii}Re_{jj} \simeq R_{\mathcal{D}_1}$. □

This result is also valid for rings which are not F-algebras but are such that there does not exist an infinite descending family $I_1 \supset I_2 \supset I_3 \supset \ldots$ of distinct left ideals (***Artinian rings***).

The converse is left as an exercise.

EXERCISES

Level 1

1. Show that the quaternions are a division ring.
2. Consider the ring of rational functions

$$\frac{Q_1(x)}{Q_2(x)}$$

over $\mathbf{C}$ where x is a variable. Show this is a field.

3. Show it is an infinite dimensional division ring over $\mathbf{C}$.
4. If z is in the center of R, prove zR is a two-sided ideal.

5. If $\mathcal{I}$ is a maximal proper ideal of $\mathcal{R}$, prove $\mathcal{R}/\mathcal{I}$ is an integral domain.
6. If e_{ii} is an idempotent in $\mathcal{R}$, prove $\mathcal{R}$ is the direct sum of $e_{ii}\mathcal{R}$ and $(1 - e_{ii})\mathcal{R}$.

Level 2

1. In the ring of $M_n(F)$:
   (a) Show all left ideals are principal (have the form $\mathcal{R}a$).
   (b) Classify left ideals in terms of the image space of a matrix on row vectors.
   (c) Classify right ideals.
   (d) Show no two-sided ideals exist.

Level 3

1. Carry through Level 2 for matrices over $\mathcal{R}_{\mathcal{D}}$.
2. Prove that a finite dimensional integral domain of **F** is a division ring.
3. If u is an element of a finite dimensional F-algebra $\mathcal{A}$, prove that there exists a polynomial in u such that $p(u)u^k = u^k$ for some k. Assume $u\mathcal{R} = u^2\mathcal{R}$. By induction show $u^k\mathcal{R} = u\mathcal{R}$. Show $p(u)u = u, p(u)$ is idempotent, and generates $u\mathcal{R}$.

CHAPTER 5

Group representations

A ***group representation*** is a homomorphism from an abstract group G to a group of $n \times n$ matrices. Two representations are equivalent if there is a similarity transformation $X \to AXA^{-1}$ changing one to another. Suppose the field **F** contains the rationals, and the group G is finite (or compact). Then it can be shown there are only a finite number of equivalence classes of representations of any fixed degree n, and these can be completely classified by the characters, i.e. the traces of the matrices assigned to each element $g \in G$.

To obtain this result, we first study the ***group ring*** $\mathsf{F(G)}$ of a group G. The group ring consists of all 'formal sums' $f_1 g_1 + f_2 g_2 + \ldots + f_n g_n$ of elements of G times coefficients in **F**. Such sums are added termwise $(e + 3g) + (2e + 5g) = 3e + 8g$. They are multiplied termwise using the products in G (commutativity does not necessarily hold). Every group representation of the group G extends to a ring representation of the ring $\mathsf{F(G)}$, and conversely. It follows from the Wedderburn theorems (Theorems 4.5.2, and 4.5.3) of Chapter 4 if $\mathbf{F} = \mathbf{C}$ that $\mathsf{F(G)}$ has a simple structure as a ring: it is a direct sum of rings $M_k(F)$ consisting of all $k \times k$ matrices over **F**, for varying k.

Over the real or complex numbers, we show that any representation is equivalent to a representation by geometrical symmetries, that is, orthogonal or unitary matrices.

A representation of $\mathsf{F(G)}$ can be formalized as a module. A module M over any ring R is a generalization of the idea of vector space to rings R which are not fields. That is M is an abelian group provided with a multiplication $R \times M \to M$ satisfying associative, distributive, and identity laws. The theory of modules is very important in more advanced abstract algebra. Two modules can be 'added' by the operation of direct sum. There is also a multiplication operation. The set of homomorphisms from one module to another is an abelian group. The kernel and image of a homomorphism are also modules.

Finite dimensional modules over $\mathsf{F(G)}$ have the property that every module is a direct sum of irreducible modules, i.e. modules with no nonzero proper submodule. Every irreducible module is isomorphic to a submodule of the group ring $F(G)$ regarded as a module over itself. Therefore to obtain all irreducible representations it suffices to find all irreducible submodules of the group ring.

Then we introduce characters, the traces of the matrices of a representation. The characters of distinct irreducible complex representations turn out to be orthonormal vectors. This is a consequence of the fact that there exists no nonzero homomorphism from one irreducible module to another unless the two are isomorphic.

It follows that every equivalence class of complex representations is completely determined by its character. Characters can be added and multiplied, and have a number of other properties.

Group representations are important in the study of systems having symmetry, for instance in physics. Frequently such systems can be decomposed in terms of the distinct irreducible representations of the system.

In the last section we discuss tensor products $M \otimes N$ of the modules. These are constructed by generators and relations expressing bilinearity. They explain the properties of ***Kronecker products*** of matrices, and enable new representations of a group to be found. Under some unproved assumptions, we find the character ring of the n-dimensional unitary group.

5.1 THE GROUP RING AND REPRESENTATIONS

For any group G and ring R there exists a ring $R(G)$ closely connected with the structure of the group.

DEFINITION 5.1.1. For a group G and ring R the *group ring* $R(G)$ is the set of all functions $f\colon G \to R$ which are zero on all but a finite number of elements of G. Such functions are added by the usual functional addition,

$$(f + h)(x) = f(x) + h(x)$$

Products are defined by

$$(f \bigcirc h)(x) = \sum_{yz=x} f(y)h(z)$$

Elements of the group ring are written as

$$r_1 g_1 + r_2 g_2 + \ldots + r_n g_n .$$

Two such sums are added and multiplied like algebraic expressions in which the g_i are treated as variables and the r_i are treated as coefficients. The coefficient r_i of g_i is interpreted as $f(g_i)$ where f is the function described in the definition. However, unless x, y commute in G, xy must be distinguished from yx in multiplication.

The description in terms of formal sums is equivalent to the description by functions: to the sum $\Sigma\, r_i g_i$ corresponds the function f such that $f(g_i) = r_i$ and

$f(x) = 0$ for other elements of g and conversely. Let sums $\Sigma\, r_i g_i$ and $\Sigma\, s_j g_j$ correspond to functions f, g. Then $(\Sigma\, r_i g_i)(\Sigma\, s_j g_j)$ is the sum $\Sigma\, t_k g_k$ where

$$t_k = \sum_{g_i g_j = g_k} r_i s_j .$$

This corresponds to a function h such that $h(g_k)$

$$h(g_k) = \sum_{g_i g_j = g_k} f(g_i) g(g_j)$$

This is the product of Definition 5.1.1.

EXAMPLE 5.1.1. Let G be the symmetric group of degree 3, with six elements e, y, y^2, x, xy, xy^2 where $y^3 = e$, $x^2 = e$, $xy = y^2 x$. In R(G),

$$(2e + 3y + 4xy) + (-e + x + 7xy) = e + 3y + x + 11xy$$

$$(2e + y + xy)(3e - 4x) = (6e + 3y + 3xy) - (8x + 4yx + 4xyx)$$

$$= 6e + 3y + 3xy - 8x - 4xy^2 - 4y^2$$

THEOREM 5.1.1. *The group ring of a group is a ring.*

Proof. We prove one of the distributive laws and the associative law of multiplication.

$$(f \bigcirc (r + s))(x) = \sum_{yz = x} f(y)(r + s)(z) = \sum_{yz = x} f(y)(r(z) + s(z))$$

$$= \sum_{yz = x} f(y) r(z) + \sum_{yz = x} f(y) s(z)$$

Therefore

$$f \bigcirc (r + s) = (f \bigcirc r) + (f \bigcirc h)$$

It follows that

$$(f \bigcirc (r \bigcirc s))(x) = \sum_{yz = x} f(y)(r \bigcirc s)(z) = \sum_{yz = x} f(y) \sum_{uv = z} r(u) s(v)$$

$$= \sum_{yuv = x} f(y) r(u) s(v)$$

$$((f \bigcirc r) \bigcirc s)(x) = \sum_{wv = x} (f \bigcirc r)(w) s(v) = \sum_{wv = x} \sum_{yu = w} f(y) r(u) s(v)$$

$$= \sum_{yuv = x} f(y) r(u) s(v)$$

The proofs of the other properties are similar. □

The group ring here will be mainly used to study group representations.

DEFINITION 5.1.2. A *representation* of a group G in a ring $\mathcal{R}$ with unit is a homomorphism h from G into the group $GL(n, R)$ of $n \times n$ invertible matrices over $\mathcal{R}$, for some $n \in \mathbf{Z}^+$. Two representations r, s are *equivalent* if and only if there exists an invertible matrix A such that $r(g) = As(g)A^{-1}$ for all $g \in G$. The number n is the *dimension* of the representation.

It follows that $h(e)$ must be the identity matrix.

A representation is thus an assignment of a matrix M_g to every group element in such a way that $M_{gh} = M_g M_h$. For a field $\mathbf{F}$ this means that a group G acts on the space $\mathbf{F}^n$ as a group of linear mappings.

EXAMPLE 5.1.2. The trivial representation is the representation $h(g) = I$ for every $g \in G$ where I is an identity matrix.

EXAMPLE 5.1.3. The cyclic group $\mathbf{Z_m}$ acts on 2-dimensional space as a group of rotations by multiples of the angle $\frac{2\pi}{m}$. If x is a generator, we have a representation sending x^j to the matrix

$$\begin{bmatrix} \cos\frac{2\pi j}{m} & \sin\frac{2\pi j}{m} \\ -\sin\frac{2\pi j}{m} & \cos\frac{2\pi j}{m} \end{bmatrix}$$

EXAMPLE 5.1.4. Any regular solid in 3-dimensional space has a finite group of symmetries G. Then G has a 3-dimensional representation as the matrices of the rotations and reflections involved. This gives, for example, a 3-dimensional representation of the alternating group of degree 5 by rotations of an icosahedron.

If $R \subset S$ a representation over R is a representation over S also. We next prove that every representation of a finite group over the real or complex numbers is equivalent to a unitary representation, that is, a group or rotations or reflections. That is, every real representation is equivalent to a homomorphism into a group of symmetries of n-dimensional space.

THEOREM 5.1.2. *Let r be a representation over* $\mathbf{R}$ *or* $\mathbf{C}$. *Then r is equivalent to a unitary representation.*

Proof. Let $f(x, y)$ denote the function on $\mathbf{C}^n$ given by

$$\sum_{g \in G} r(g)x \cdot r(g)y$$

Then

$$f(r(h)x, r(h)y) = \sum_g r(g)r(h)x \cdot r(g)r(h)y = \sum_{gh} r(gh)x \cdot r(gh)y$$
$$= \sum_g r(g)x \cdot r(g)y = f(x, y)$$

Here, since G is a group, gh ranges over all elements of G if g does. This shows f is invariant.

We show that $f(x, y)$ is what is called a ***nonsingular bilinear Hermitian form.*** We have

$$f(x, y) = \overline{f(y, x)}$$

$$f(x + y, z) = \Sigma\, r(g)(x + y) \cdot r(g)z = \Sigma\, r(g)x \cdot r(g)z + \Sigma\, r(g)y \cdot r(g)z$$
$$= f(x, z) + f(y, z)$$

$$f(ax, z) = \Sigma\, r(g)ax \cdot g(z) = af(x, z)$$

These imply $f(z, x + y) = f(z, x) + f(z, y)$ and $f(x, az) = a^* f(x, z)$.

In addition

$$f(x, x) = \sum_{g \in G} r(g)x \cdot r(g)x > 0$$

for all $x \neq 0$.

By the ***Gram-Schmidt Process*** we can construct a new basis $v_1, v_2, \ldots, v_n$ such that $f(v_i, v_i) = 1$ and $f(v_i, v_j) = 0$ if $i \neq j$. To do this let $u_1, u_2, \ldots, u_n$ be any basis. Let

$$v_1 = \frac{u_1}{\sqrt{f(u_1, u_1)}}$$

so $v_1 \cdot v_1 = 1$. For $i = 1$ to n, let

$$w_i = u_i - \sum_{j<i} f(u_i, v_j)v_j$$

Then $(w_i \cdot v_j) = 0$ for $j < i$. Let

$$v_i = \frac{w_i}{\sqrt{f(w_i, w_i)}}$$

Then v_i is the required basis.

Now we claim that with respect to this new basis, the matrix X of $r(h)$ will be unitary. Let $r(h)v_i = \Sigma\, x_{ij}v_j$. Then by invariance $f(v_i, v_k) = \delta_{ik} = f(r(h)v_i, r(h)v_k)$ where $\delta_{ik} = 1$ if $i = k$ and $\delta_{ik} = 0$ if $i \neq k$. This gives

$$\delta_{ik} = f(\Sigma\, x_{ij}v_j, \Sigma\, x_{km}v_m) = \Sigma\, x_{ij}\overline{x}_{km} f(v_j, v_m) = \Sigma\, x_{ij}\overline{x}_{km}\delta_{jm}$$
$$= \Sigma\, x_{im}\overline{x}_{km}$$

Therefore $X(X^T)^* = I$. Therefore X is unitary. □

This result, by a similar proof, holds for representations of compact topological groups.

One relationship between representations and the group ring is that there is a 1-1 correspondence between group homomorphisms from G to $n \times n$ invertible matrices over $\mathbf{R}$ and ring homomorphisms from R(G) to the ring of $n \times n$ matrices over $\mathbf{R}$, sending e to I, provided $\mathbf{R}$ is a commutative ring with unit. Here e denotes the identity element of G and I denotes the identity matrix respectively. This mapping is defined by sending $\Sigma\, r_g g$ to $\Sigma\, r_g h(g)$.

EXERCISES

Level 1

1. Compute in the group ring of the symmetric group of degree 3 over $\mathbf{Z}$:
   (a) $(2e + 6x + 5y + 7y^2) + (2e + 10x + xy + xy^2)$
   (b) $(e + y + y^2)(y)$
   (c) $(e + y + y^2)(x)$
   (d) $(e + y + y^2)(2x + 3y)$
2. Define a representation of $\mathbf{Z}_2 \times \mathbf{Z}_2$ into 2×2 diagonal matrices over $\mathbf{Z}$. Generalize this.
3. Define a representation of the symmetric group into $n \times n$ permutation matrices over $\mathbf{Z}$.

Level 2

1. Prove that in a group ring F(G) there always exist nonzero elements which have no inverses, if $|G| > 1$.
2. Show $\mathrm{Q}(\mathrm{Z}_2)$ is isomorphic to $\mathbf{Q} \oplus \mathbf{Q}$.
3. What are the nonzero ideals of $\mathrm{Q}(\mathrm{Z}_2)$?
4. Show that

$$m \to \begin{bmatrix} 1 & 0 \\ m & 1 \end{bmatrix}$$

   is a representation of $\mathbf{Z}$ not equivalent to a unitary representation.
5. Find a real orthogonal representation of the symmetric group of degree 3 equivalent to this one:

$$x \to \begin{bmatrix} 0 & 1 \\ 1 & 0 \end{bmatrix}, \quad y \to \begin{bmatrix} 0 & 1 \\ -1 & -1 \end{bmatrix}$$

Level 3

1. Consider the rational group ring $\mathrm{Q}(\mathrm{Z}_m)$. Define homomorphisms $h: \mathrm{Q}(\mathrm{Z}_m) \to \mathbf{Q}$ by $h(x) = 1$, $x \in \mathbf{Z}_m$ and $\mathrm{Q}(\mathrm{Z}_m) \to \mathbf{F}$ where $\mathbf{F}$ is the subfield of $\mathbf{C}$ generated by

$$\omega = e^{\frac{2\pi i}{m}}$$

If m is prime the number ω satisfies $\omega^{m-1} + \omega^{m-2} + \ldots + \omega + 1 = 0$ but no equation of lesser degree over **Q**. Use this to show the map into $\mathbf{Q} \oplus \mathbf{F}$ is an isomorphism.

2. Prove that a $n \times n$ matrix A over **Q** is a linear combination of permutation matrices if and only if the vector $(1, 1, \ldots, 1)$ is both a row and a column eigenvector of A.
3. Show $Q(Z_4)$ is isomorphic to $\mathbf{Q} \oplus \mathbf{Q} \oplus \mathbf{F}$ where **F** is the set of numbers $\{a + bi : a, b \in \mathbf{Q}\}$.
4. Show that a group of order n has a nontrivial representation of degree n.
5. Show that any group $\mathbf{Z_m}$ has a nontrivial 1-dimensional complex representation h. Show the real 2-dimensional representation of the example is equivalent over **C** to the direct sum representation:

$$x \to \begin{bmatrix} h(x) & 0 \\ 0 & h^*(x) \end{bmatrix}$$

5.2 MODULES AND REPRESENTATIONS

We next show that a representation over **F** with unit is equivalent to a finite dimensional module over F(G). A module can be described as a ring acting linearly on an abelian group. Modules play an important role in group representation theory, algebraic topology, and other areas of modern mathemetics.

DEFINITION 5.2.1. A *left* (*right*) *module* over a ring R is an abelian group M together with a function $R \times M \to M$ ($M \times R \to M$) such that (1) $(r + s)m = rm + sm$ (respectively $m(r + s) = mr + ms$), (2) $r(m + n) = rm + rn$ (respectively $(m + n)r = mr + nr$), (3) $(rs)(m) = r(sm)$ (respectively $m(sr) = (ms)r$), (4) $(0)m = 0$ (respectively $m(0) = 0$).

EXAMPLE 5.2.1. A set which is a left module over itself is a ring.

EXAMPLE 5.2.2. The n-fold direct sum $\mathbf{R} \oplus \mathbf{R} \oplus \ldots \oplus \mathbf{R}$ is a left module by $s(r_1, r_2, \ldots, r_n) = (sr_1, sr_2, \ldots, sr_n)$. A module isomorphic to such a direct sum is called a ***free module***.

DEFINITION 5.2.2. A *bimodule* over a ring is a module which is simultaneously a left module and a right module such that $(rm)s = r(ms)$.

EXAMPLE 5.2.3. Over a commutative ring every left or right module is a bimodule by the definition $rm = mr$.

EXAMPLE 5.2.4. Any free module $\mathbf{R} \oplus \mathbf{R} \oplus \ldots \oplus \mathbf{R}$ is a bimodule.

For noncommutative rings it is necessary to distinguish right and left modules.

DEFINITION 5.2.3. A module is *unitary* if and only if $1 \cdot m = m$ $(m \cdot 1 = m)$ for all $m \in M$, where 1 is the identity over R.

For the remainder of this chapter, all modules will be unitary.

EXAMPLE 5.2.5. Any vector space is a unitary module over the field in question.

DEFINITION 5.2.4. A *homomorphism* of left (right) R-modules from M to N is a function f from M to N such that $f(x+y) = f(x) + f(y)$ and $f(rx) = rf(x)$ $(f(xr) = f(x)r)$ for all $x, y \in M$, $r \in R$. A 1-1 onto homomorphism of modules is an isomorphism.

EXAMPLE 5.2.6. Any two vector spaces of the same dimension are isomorphic $\mathbf{F}$-modules.

Every unitary $F(G)$-module is a vector space over $\mathbf{F}$. We will call it finite dimensional if it is so as a vector space over $\mathbf{F}$, and its dimension is the vector space dimension.

THEOREM 5.2.1. *There exists a 1-1 correspondence between equivalence classes of representations of a group G over* $\mathbf{F}$ *with unit and isomorphism classes of unitary, finite dimensional left* $F(G)$*-modules.*

Proof. To every representation h of G over $\mathbf{F}$ we make the vector space $\mathbf{F}^n$ into a G-module by setting $(\Sigma\, c_g g)v = \Sigma\, c_g h(g)v$. Conversely for any finitely generated $F(G)$-module, choose a basis $m_1, m_2, \ldots, m_n$ for M. Multiplication by g defines a linear transformation f_g on M since $g(m+n) = gm + gn$ and $g(an) = (ga)n = ag(n)$. These linear transformations satisfy $f_{gh}(m) = gh(m) = g(h(m)) = f_g f_h(m)$. So $f_{gh} = f_g f_h$. Let $h(g)$ be the matrix of f_g. Then $h(g)h(g^{-1}) = h(e) = I$ so $h(g)$ is invertible. And $h(r)h(s) = h(rs)$. This gives a representation. Different choices of a basis change each $h(g)$ to a matrix $Xh(g)X^{-1}$ so the representations are equivalent. Isomorphic modules likewise give equivalent representations and equivalent representations give isomorphic modules.

If we choose the standard basis for $\mathbf{F}^n$ then the matrix representation $h(g)$ goes to the module $h(g)v$ which again yields $h(g)$ as matrix. Conversely if we go from a module M with basis $v_1, v_2, \ldots, v_n$ to a matrix $h(g)$ then $h(g)v$ is the original gv. So $\Sigma\, c_g h(g)v = \Sigma\, c_g gv$. This establishes that the two correspondences are inverse to one another. Therefore each is 1-1. □

Right modules could have been used in Theorem 5.2.1 since to every left module over $F(G)$ there corresponds a right module with multiplication given by $m\,(\Sigma\, a_i g_i) = \Sigma\, a_i g_i^{-1} m$.

So it suffices to study F(G)-modules to determine equivalence classes of representations. We will show next that every finite dimensional F(G)-module is a sum of irreducible F(G)-modules.

DEFINITION 5.2.5. The *direct sum* of the left (right) R-modules $M_1, M_2, \ldots, M_n$ is the Cartesian product $M_1 \oplus M_2 \oplus \ldots \oplus M_n$ with operations defined by $(x_1, x_2, \ldots, x_n) + (y_1, y_2, \ldots, y_n) = (x_1 + y_1, x_2 + y_2, \ldots, x_n + y_n)$ and $c(x_1, x_2, \ldots, x_n) = (cx_1, cx_2, \ldots, cx_n)$ (respectively $(x_1, x_2, \ldots, x_n)c = (x_1 c, x_2 c, \ldots, x_n c)$).

EXAMPLE 5.2.7. For vector spaces this is the same as the previous definition of direct sum.

DEFINITION 5.2.6. An additive subgroup N of a left (right) R-module M is a *submodule* if and only if for all $r \in R$, $n \in N$ we have $rn \in N$ (respectively $nr \in N$).

EXAMPLE 5.2.8. An ideal is a submodule of a ring over itself.

EXAMPLE 5.2.9. For vector spaces, a submodule is a subspace.

EXAMPLE 5.2.10. The subsets $\{0\}$, M submodules of any module M.

DEFINITION 5.2.7. A nonzero module M is *irreducible* if and only if it has no submodule except M, $\{0\}$.

EXAMPLE 5.2.11. A vector space **V** is irreducible as a **F**-module if and only if it is 1-dimensional.

DEFINITION 5.2.8. For every submodule $N \subset M$, there exists a *congruence relation* on M defined by $r \sim s$ if and only if $r - s \in N$. The set of equivalence classes forms a module with operations $\bar{x} + \bar{y} = \overline{x + y}$, $r\bar{x} = \overline{rx}$ (or $\bar{x}r = \overline{xr}$). This is called the *quotient module* M/N.

THEOREM 5.2.2. *Let G be a finite group and let* **F** *be* **C** *or* **R**. *For every submodule N of a finite dimensional* F(G)*-module M there exists a complementary submodule T such that $M \simeq N \oplus T$. Moreover $N/M \simeq T$.*

Proof. By Theorem 5.1.2, we may assume G acts on M by unitary matrices. Let T be the space of vectors orthogonal to all vectors in N. Then T is a submodule since if $s \in T$, $x \in N$, $\Sigma\, c_g g \in$ F(G) we have

$$\Sigma\, c_g g s \cdot n = \Sigma\, c_g(s \cdot g^{-1} n) = \Sigma\, c_g(0) = 0\,.$$

Here the g^{-1} comes from the unitary property and $g^{-1}n \in N$ so $s \cdot g^{-1}n = 0$. Moreover $T \cap N = \{0\}$ and $T + N = M$ as vector spaces. This implies the mapping $(s, n) \to s + n$ is an isomorphism from $T + N$ into M.

The mapping $s \to \bar{s}$ from T to N/M has kernel 0 and is onto, therefore it is an isomorphism. □

COROLLARY 5.2.3. *Every finite dimensional* F(G)*-module is a direct sum of irreducible modules.*

Proof. We repeatedly express a module M as a direct sum until we arrive at modules having no nonzero submodules. If the dimension is n this process must end after at most $(n - 1)$ steps. □

DEFINITION 5.2.9. The *characteristic of a field* **F** (char (**F**)) is the least positive integer n such that $n1 = 1 + 1 + \ldots + 1 = 0$ if such an n exists. Else it is zero.

EXAMPLE 5.2.12. char $(\mathbf{Q}) = 0$, char $(\mathbf{Z_p}) = p$.

Theorem 5.2.2 is actually true for any field **F** such that (char (**F**), $|G|$) = 1. A proof is given in the exercises (see Level 3, Exercises 2-5).

EXERCISES

Level 1

1. Explain why every commutative group is a **Z**-module.
2. If f is a homomorphism from a ring R to a ring S explain how S can be regarded as an R-module.
3. Prove that **Z** is not a unitary **Q**-module.
4. Prove that the set of n-dimensional vectors over **F** is a bimodule over the ring $M_n(F)$.
5. For $a \in R$, prove that aM is a submodule of M provided a commutes with all elements of R.

Level 2

1. Suppose a group G acts by permutations on a set X. Find a **Z**-module associated with this action. (It can be taken as the set of functions $X \to \mathbf{Z}$.)
2. Prove that if M is a module over **R**, every invertible element of the centre of **R**, $\{x : xy = yx \text{ for all } y \in \mathbf{R}\}$ gives a module isomorphism $M \to M$.
3. Let M^n be the direct sum of n copies of an **R**-module M. Prove M^n is a module over the ring of $n \times n$ matrices over **R**.
4. Show that the subgroup of $\mathbf{Z} \times \mathbf{Z}$ generated by $\{(1, 1), (2, 3)\}$ is itself a free **Z**-module.

5. Do the same for the subgroup generated by $\{(1,4),(2,2),(4,1)\}$.
6. Prove that every abelian group of order m is a $\mathbf{Z_m}$-module.

Level 3

1. An integral domain $\mathcal{D}$ is called a ***principal ideal domain*** if and only if every ideal has the form $\mathcal{D}(a)$ for $a \in \mathcal{D}$. Prove every finitely generated module over a principal ideal domain is isomorphic to a direct sum of modules of the form $\mathcal{D}/\mathcal{D}(a)$.
2. If $\mathbf{Z_r} \not\subset \mathbf{F}$ for $r \,|\, |G|$ every $\mathrm{F(G)}$-submodule N of an $\mathrm{F(G)}$-module M is a direct summand. Let $\mathbf{V} \subset M$ be a vector space complement of N in M, and let $h : M \to \mathbf{V}$ be a projection mapping onto $\mathbf{V}$ with kernel N. Let

$$h_1(x) = \frac{1}{|G|} \sum_G g h(g^{-1}(x))$$

   Prove h_1 is a module homomorphism from M to M. The property $\mathbf{Z_r} \not\subset \mathbf{F}$ guarantees $|G|$ has an inverse in $\mathbf{F}$.
3. Let N and h_1 be the same as in the above exercise. Prove $N \subset \ker(h_1)$.
4. Prove that the mapping $\mathbf{V} \to M^{h_1} \to M \to M/N \simeq \mathbf{V}$ is the identity.
5. Prove image of h_1 is a complement of N.

5.3 IRREDUCIBLE REPRESENTATIONS

We have established that every finite dimensional representation of a finite group over $\mathbf{R}$ or $\mathbf{C}$ is a direct sum of irreducible representations. In this section we show the number of irreducible representations is finite, and every irreducible representation is a submodule of the group module $\mathbf{R}$ itself. We further study the decomposition of the group ring as a direct sum, the number of occurrences of each irreducible module, and homomorphisms among irreducible modules over $\mathbf{C}$.

THEOREM 5.3.1. *Every irreducible module M over a ring R has the form R/K where K is a maximal proper left ideal.*

Proof. Let $x \in M$, $x \neq 0$. Then Rx is a submodule of M, so $Rx = M$. We have a module homomorphism $R \to Rx = M$ sending 1 to x. If K is the kernel of this mapping, $R/K \simeq M$ and K is a left-ideal. If $K \subset I$ for another proper ideal I then I/K would be a nonzero proper submodule of M/K. Therefore K is maximal. □

In matrix terms, an irreducible representation is one which is not equivalent to a representation by block lower triangular matrices

$$\begin{bmatrix} * & 0 \\ * & * \end{bmatrix}$$

DEFINITION 5.3.1. The *regular matrix representation* of a finite group $G = \{g_1, g_2, \ldots, g_n\}$ is the representation by permutation matrices $p_{ij}(g)$ where $p_{ij}(g) = 1$ if and only if $g_i g = g_j$.

These are the matrices of the regular representation by permutations.

EXAMPLE 5.3.1. For the group $\mathbf{Z}_2$ the matrices are

$$\begin{bmatrix} 1 & 0 \\ 0 & 1 \end{bmatrix}, \begin{bmatrix} 0 & 1 \\ 1 & 0 \end{bmatrix}$$

It can be verified that the corresponding module is precisely the group ring. An isomorphism is given by sending the basis element g_i for the group ring to the $(0, 1)$-vector whose only 1 is in place i. The regular representation has dimension $|G|$.

For the remainder of this chapter we will be concerned with representations over the field of complex numbers, although the next two theorems are valid for any field $\mathbf{F}$ such that $(\text{char}(\mathbf{F}), |G|) = 1$.

LEMMA 5.3.2. (Schur's Lemma.) *Let* M, N *be irreducible modules over* $C(G)$. *If* M *and* N *are not isomorphic then every homomorphism from* M *to* N *is zero. If* M, N *are isomorphic then every nonzero homomorphism is an isomorphism.*

Proof. Consider $f: M \to N$. Suppose f is nonzero. Then the image of f is a nonzero submodule of N so it must be N. The kernel of f is a proper submodule of M so it must be 0. So f is an isomorphism. □

THEOREM 5.3.3. *Let the regular representation of* $C(G)$ *be decomposed as a direct sum of irreducible modules* M_i. *Every irreducible module* M *is isomorphic to some* M_i.

Proof. Let y be any nonzero element of M. We have a module homomorphism from $C(G) = \oplus M_i$ to M sending $1 \in C(G)$ to y and $\Sigma\, c_g g$ to $\Sigma\, c_g g y$. For some i this mapping must be nonzero on M_i. By Lemma 5.3.2 M_i is isomorphic to M. □

COROLLARY 5.3.4. *A group of order n has at most n inequivalent irreducible representations.*

So it suffices to study the regular representation to obtain a complete set of irreducible representations. But the structure of the regular representation is in effect known from the last section of Chapter 4. The ring $C(G)$ is a direct sum of matrix rings $M_n(C)$ of $n \times n$ matrices over $\mathbf{C}$ and these matrix rings correspond to distinct idempotents, hence we have the following theorem.

THEOREM 5.3.5. *An irreducible representation of dimension n occurs precisely n times in the regular representation.*

Proof. Without going into detail, each matrix algebra $M_n(C)$ of $n \times n$ matrices over **C** yields precisely n copies of a single irreducible representation, corresponding to the n columns of each matrix. Distinct matrix algebras give distinct representations. The representation of $M_n(C)$ on the set of column vectors can be verified to be irreducible. □

PROPOSITION 5.3.6. *For any irreducible representation $\mathcal{M}$, the vector space of homomorphisms $\mathcal{M}$ to $\mathcal{M}$ is isomorphic to* **C**.

Proof. We show the module **V** of row vectors over $M_n(C)$ has this property. Here $M_n(C)$ denotes the $n \times n$ matrices over **C**. Let $f: \mathbf{V} \to \mathbf{V}$ be an isomorphism such that $f(Av) = Af(v)$ for every v. Let A be a $(0, 1)$-matrix with exactly one 1 in the (i, i)-entry. Let e_i be the $(0, 1)$-vector having exactly one 1, in place i. Then $Af(v) = f(A(v)) = f(v_i e_i) = v_i f(e_i)$. Therefore the i component of $f(v)$ which is that of $Af(v)$ is $v_i f(e_i)_i$. So f is a diagonal matrix D with entries $f(e_i)_i$. Let A have $a_{ij} = a_{ij} = 1$ and all other entries zero. Then $AD = DA$ implies $d_{ii} = d_{jj}$ by looking at the (i, j)-entry of each. □

EXERCISES

Level 1

1. If the group is commutative show the group ring is also.
2. For a commutative group if C(G) is isomorphic to a direct sum of $M_n(C)$ explain why no n can exceed 1. Therefore all irreducible representations are 1-dimensional.
3. Show that n distinct irreducible representations of $\mathbf{Z_n}$ are obtained by sending a generator 1 to $\frac{2\pi ik}{n}$. Therefore these are precisely the irreducible representations of $\mathbf{Z_n}$.
4. For any irreducible representation of a group H into $M_n(C)$ and for any homomorphism $f: G \to H$ which is onto prove the representation $f: G \to M_n(C)$ is irreducible.
5. Use Exercise 4 to obtain p^2 irreducible representations of $\mathbf{Z_p} \times \mathbf{Z_p}$.

Level 2

1. How many times does the regular representation contain the trivial representation?
2. Identify the trivial representation as a submodule of $\mathbf{Z_n}$.
3. For the symmetric group give a linear representation of the form $S_n \to \mathbf{Z_2} \to M_n(C)$.

4. Prove the symmetric group of order 3 has a 2-dimensional representation given by symmetries of an equilateral triangle.
5. Show any 1-dimensional representation is commutative. Hence a 2-dimensional representation by non-commuting matrices must be irreducible.
6. Using the above 5 exercises give the complete set of irreducible representations of the symmetric group of degree 3.

Level 3

1. Generalize Level 2 to obtain all irreducible representations of the dihedral group of order $2n$, n odd. (They are all 1 or 2-dimensional.) Note that if the group is written in terms of generators and relations $x, y: x^2 = y^n = e$, $xy = y^{-1}x$ that there is a quotient map on to $\mathbf{Z}_2$ sending y to e where e is an identity and there are automorphisms $y \to y^k$ for any relatively prime to n. These with symmetries of a regular n-gon give the irreducible representations.
2. Give the irreducible representations of the dihedral group for n even. (There are additional 1-dimensional representations sending $y^2 \to e$ but not y.)

5.4 GROUP CHARACTERS

A ***character*** χ of a representation assigns a complex number to each element of the representation. A character identifies a representation completely. Characters have many properties: $\chi(M \oplus N) = \chi(M) + \chi(N)$, $\chi(M \otimes N) = \chi(M)\chi(N)$ where $\otimes$ is an operation called ***tensor product***. Thus if one adds negative values they form a ring. The characters of distinct irreducible representations are orthogonal.

DEFINITION 5.4.1. Let $r : G \to M_n(C)$ be a representation where $M_n(C)$ denotes the set of all $n \times n$ matrices over $\mathbf{C}$. Then the *character* associated with r is $\chi(g) = \mathrm{Tr}\,(r(g))$.

EXAMPLE 5.4.1. Consider this representation of $\mathbf{Z}_2$:

$$e = \begin{bmatrix} 1 & 0 \\ 0 & 1 \end{bmatrix}, \quad x = \begin{bmatrix} 0 & 1 \\ 1 & 0 \end{bmatrix}$$

The character is $\chi(e) = 2$, $\chi(x) = 0$.

PROPOSITION 5.4.1. *The character of a direct sum of two representations is the sum of the characters of the separate representations.*

Proof. For a direct sum of representations r_1, r_2 the matrices look like this:

$$\begin{bmatrix} r_1(g) & 0 \\ 0 & r_2(g) \end{bmatrix}$$

The trace of such a matrix, the sum of the main diagonal entries, is $\mathrm{Tr}\,(r_1(g)) + \mathrm{Tr}\,(r_2(g))$. □

We have briefly mentioned before the Kronecker product, sending an $m \times m$ matrix A and an $n \times n$ matrix B to an $nm \times nm$ matrix $A \boxtimes B$ such that $(A \boxtimes B)_{in+j-n,\, un+v-n} = a_{iu} b_{jv}$. This operation is multiplicative. $(A \boxtimes B)(C \boxtimes D) = AC \boxtimes BD$. It follows that the Kronecker product of two representations gives a new representation.

PROPOSITION 5.4.2. *The character of a Kronecker product of two representations is the product of the characters of the representations.*

Proof. We need to show that $\mathrm{Tr}(A \boxtimes B) = \mathrm{Tr}(A)\,\mathrm{Tr}(B)$. Write A, B in subtriangular form with eigenvalues on the main diagonal, by taking a similarity transformation. Then $A \otimes B$ will also be in subtriangular form, and its eigenvalues are the products $a_{ii} b_{jj}$. So $\mathrm{Tr}(AB) = \sum_{i,j} a_{ii} b_{jj} = \sum_i a_{ii} \sum_j b_{jj} = \mathrm{Tr}(A)\,\mathrm{Tr}(B)$. □

In the following proof we use the fact that if r, s are representations, and Y is a matrix such that $r(g)Y = Ys(g)$ then $v \to Yv$ gives a module homomorphism f between the corresponding modules. The reason is that $f(r(g)v) = Yr(g)v = s(g)Yv = s(g)f(v)$.

THEOREM 5.4.3. *Let r, s be irreducible representations giving matrices $(r_{ij}(g), s_{ij}(g))$. Then $\Sigma\, r_{ij}(g) s_{km}(g^{-1}) = 0$ unless r is equivalent to s. If $r = s$ then this sum is 0 unless $i = m$, $j = k$ and in that case it is $\frac{|G|}{n}$ where n is the dimension of the representation.*

Proof. For any matrix B the rectangular matrix

$$Y = \Sigma\, r(g) B s(g^{-1})$$

satisfies

$$\begin{aligned} r(h)Y &= \Sigma\, r(hg) B s(g^{-1}) \\ &= \Sigma\, r(hg) B s((hg)^{-1}) s(h) \\ &= Ys(h) \end{aligned}$$

Thus it gives a module homomorphism from one representation to the other.

For r, s distinct it must be zero by Lemma 5.3.2. Let $r = s$. Then all such homomorphisms Y must be a multiple of the identity aI, by Proposition 5.3.6.

Let B be a $(0, 1)$-matrix with a 1 precisely in its (j, k)-entry. Then the sum in question is the (i, m)-entry of Y. This proves the first statement. Also if $i \neq m$ the sum is 0 since $Y = aI$, and for $i = m$ is independent of m. But $\Sigma\, r_{ij}(g) r_{km}(g^{-1}) = \Sigma\, r_{ij}(g^{-1}) r_{km}(g)$ if we let the sum run over g^{-1} instead of g. Therefore the sum for ij, km is the same as for km, ij. Therefore if $j \neq k$ it is zero, and all the nonzero sums are equal to some number a. They can be evaluated by taking $B = I$, in which case $Y = |G|I$. So $naI = |G|I$ since taking $B = I$ means we add n equal sums for $j = 1, 2, \ldots, n$. □

We observe that for a unitary matrix M, the eigenvalues of M^{-1} are conjugates of those of M. Hence $\chi(g^{-1}) = \chi(g^*)$.

DEFINITION 5.4.2. The *inner product* of two characters (χ_1, χ_2) is

$$\frac{1}{|G|}\, \Sigma\, \chi_1(g)\chi_2(g^{-1}) \;=\; \frac{1}{|G|}\, \Sigma\, \chi_1(g)\chi_2(g^*)$$

COROLLARY 5.4.4. *If χ_1, χ_2 are irreducible then $(\chi_1, \chi_2) = 0$ if χ_1, χ_2 are not equivallent, and $(\chi_1, \chi_2) = 1$ if χ_1, χ_2 are equivalent.*

Proof. This follows from Theorem 5.4.3 by summing $\Sigma\, r_{ii}(g) s_{kk}(g^{-1})$ over all i, k. There are n equal nonzero terms, with $i = k$. For a fixed g this gives $\chi_1(g)\chi_2(g^{-1})$. □

COROLLARY 5.4.5. *For any representation χ, the value (χ, χ) is 1 if and only if χ is irreducible.*

Proof. Write the character χ as $\Sigma\, n_i\chi_i$ corresponding to the irreducible representations making it up. Then

$$(\chi, \chi) \;=\; \Sigma\, n_i n_j (\chi_i, \chi_j) \;=\; \Sigma\, n_i^2 (\chi_i, \chi_i) \;=\; \Sigma\, n_i^2$$

by Theorem 5.4.3. This is 1 if and only if there is only one nonzero n_i and it is 1. □

COROLLARY 5.4.6. *Two representations with the same character are equivalent.*

Proof. Write χ as $\Sigma\, n_i\chi_i$ corresponding to the irreducible representations making up the representation. Then (χ, χ_i) gives the number of times n_i that χ_i occurs. Thus the number of copies of each irreducible summand can be determined from χ. □

Most of these results go through for compact continuous groups, if the sums are replaced by integrals.

There are many other relationships involving characters, some sketched in the exercises.

EXERCISES

Level 1

1. Prove that the number of times the trivial representation occurs in a representation with character χ is

$$\frac{1}{|G|} \Sigma\, \chi(g)\,.$$

2. What is $\chi(e)$? Here e stands for an identity.
3. Verify the relations $(\chi_i, \chi_j) = 0$ if $i \neq j$ for irreducible representations of $\mathbf{Z_p}$. Use the fact for any nontrivial pth root x of unity $x^{p-1} + x^{p-2} + \ldots + 1 = 0$.
4. Prove that the trace of any element of order p is a sum of pth roots of unity, i.e. numbers satisfying $x^p = 1$.
5. Show $\chi(g) = \chi(hgh^{-1})$ for any character χ.

Level 2

1. Write the characters of the dihedral group of order $2n$, n odd.
2. Verify the orthogonality relations for this group.
3. For an irreducible representation of degree 2 of the symmetric group of degree 3, what is $\chi \boxtimes \chi$? What irreducible representations make up the Kronecker product of this representation with itself?
4. Let C_k be $\Sigma\, g$ over the kth conjugate class π_k. Prove that $C_i\, C_j$ is a sum of elements C_k with integer coefficients, $C_i\, C_j = \Sigma\, n_{ijk}\, C_k$ for some integers n_{ijk}.
5. In an irreducible representation C_k being in the center, must go to a certain multiple of the identity aI. Let $x_k \in \pi_k$ and let the dimension be n and let $|\pi_k| = h_k$. Then $\mathrm{Tr}\,(C_k) = h_k \chi(x_k) = na$. What is then the image of C_k?
6. Let

$$y_k = \frac{h_k \chi(x_k)}{n}\,.$$

   We have from previous results $y_i\, y_j = \Sigma\, n_{ijk}\, y_k$. So each y_i satisfies a monic polynomial with integer coefficients, the characteristic polynomial of (a_{ij}) where $a_{jk} = n_{ijk}$. Such numbers are called ***algebraic integers.***
7. Show a basis for the center of C(G) is given by the sums C_k since elements in the same class must have the same coefficient. The center, as a ring, is a direct sum of copies of the complex numbers, one for each irreducible representation. Therefore the number of distinct irreducible representations is what?

Level 3

1. Try to prove the ***class orthogonality relations***

$$\frac{1}{|G|}\sum_i \chi_i(g)\chi_i(h)^* = \begin{cases} 1 \text{ for } g \text{ conjugate to } h \\ 0 \text{ otherwise} \end{cases}$$

   where the sum is taken over all irreducible representations. characters

2. Assuming sums and products of algebraic integers are algebraic integers and the results of Exercise 7 of Level 2 and the orthogonality relations of the above exercise, prove the degree of an irreducible representation divides the order of the group.

5.5 TENSOR PRODUCTS

A ***tensor product*** is a kind of multiplicative operation on vector spaces or modules. Its definition uses bilinear forms. It gives a more natural explanation of the somewhat arbitrary Kronecker product of representations.

DEFINITION 5.5.1. A function b from $M \times N$ to G is *bilinear,* for a right R-module M, a left R-module N, and an abelian group G, if and only if for all $x, y \in M$, $z, w \in N$, $r \in R$:

(1) $b(xr, z) = b(x, rz)$

(2) $b(x + y, z) = b(x, z) + b(y, z)$

(3) $b(x, z + w) = b(x, z) + b(y, z)$

EXAMPLE 5.5.1. The multiplication $b(x, y) = xy$ is bilinear in any ring R.

EXAMPLE 5.5.2. The inner product $x \cdot y$ is a bilinear mapping of vector spaces.

The tensor product is defined to be, in category-theoretical language, a universal object for bilinear mappings.

DEFINITION 5.5.2. An abelian group H provided with a bilinear map $f: M \times N \to H$ is called a *tensor product* of a left R-module N and a right R-module M if and only if for any abelian group G and bilinear map $b : M \times N \to G$ there exists a unique homomorphism $h : H \to G$ such that $f \circ h = b$.

EXAMPLE 5.5.3. The tensor product of R with a R-module M is M.

PROPOSITION 5.5.1. *Any two tensor products are isomorphic by a natural isomorphism.*

Proof. Let H_1, H_2 be tensor products of M, N. There exist bilinear mappings $b_i : M \times N \to H_i$. These by the definition, give mappings $g_1 : H_1 \to H_2$ and $g_2 : H_2 \to H_1$. Moreover $b_1 g_1 = b_2$, $b_2 g_2 = b_1$. Therefore $g_1 g_2$ is the identity on the image of b_1 and $g_2 g_1$ is the identity on the image of b_2.

We will show b_i are onto. Suppose b_1 is not onto but has image E contained in H_1. Then by the argument just given there exists a mapping $H_1 \to E$ such that $E \to H_1 \to E$ is the identity. Then for $b = f$ there exist two distinct homomorphisms:

$$h_1 = \text{identity}$$

$$h_2 = \text{the map } H_1 \to E \to H_1$$

such that $fh_i = b$. This contradicts uniqueness. Thus the b_i are onto. Thus $g_1 g_2$ and $g_2 g_1$ are the identity. So each is an isomorphism. □

The existence of a tensor product follows by a construction in terms of generators and relations. This construction will not be proved in detail.

Take a free abelian group (sum of copies of **Z**), with one generator denoted $m \otimes n$ for each pair $(m, n) \in M \times N$. Introduce relations of three types:

(1) $xr \otimes z = x \otimes rz$

(2) $(x + y) \otimes z = x \otimes y + y \otimes z$

(3) $x \otimes (z + w) = x \otimes z + x \otimes w$

The group with these generators and relations can be verified to be the tensor product by the same sort of proof as was presented for semigroups.

Tensor products have many properties.

T1. $R \otimes M \simeq M$.

T2. $(M \oplus N) \otimes G = (M \otimes G) \oplus (N \otimes G)$.

T3. If M is a bimodule, $M \underset{R}{\otimes} N$ is a left module.

T4. $(M \underset{R}{\otimes} N) \underset{R}{\otimes} G \simeq M \underset{R}{\otimes} (N \underset{R}{\otimes} G)$, if these are defined.

There are dual properties where left, right factors are interchanged.

In group representation theory, the tensor product is taken usually not over F(G) but over **F** only, that is, it is a tensor product of vector spaces.

THEOREM 5.5.2. *Let $v_1, v_2, \ldots, v_n$ be a basis for a vector space* **V** *and $w_1, w_2, \ldots, w_n$ be a basis for vector space* **W**. *Then $v_i \otimes w_j$ is a basis for* $\mathbf{V} \otimes \mathbf{W}$ *where $i, j = 1, 2, \ldots, n$.*

Proof. The set $x \otimes y$ spans $\mathbf{V} \otimes \mathbf{W}$ by construction, for $x \in \mathbf{V}$, $y \in \mathbf{W}$. Write $x = \Sigma f_i v_i$, $y = \Sigma g_i w_i$. Then

$$x \otimes y = \Sigma f_i v_i \otimes \Sigma g_j w_j = \sum_i f_i (v_i \otimes \Sigma g_j w_j) = \sum_i f_i g_j (v_i \otimes w_j)$$

by repeated applications of the defining relations of the construction. Therefore $v_i \otimes w_j$ form a spanning set.

Take a basis denoted u_{ij} for a space $\mathbf{U}$ of dimension $(\dim \mathbf{V})(\dim \mathbf{W})$. Then there is an onto bilinear map from $\mathbf{V} \otimes \mathbf{W}$ to $\mathbf{U}$ defined by $b(\Sigma f_i v_i, \Sigma g_i w_i) = (\Sigma f_i g_j u_{ij})$. The elements $v_i \otimes w_j$ go to linearly independent elements. Therefore they are linearly independent in $\mathbf{V} \otimes \mathbf{W}$. □

If a group G acts on $\mathbf{V}, \mathbf{W}$ then G acts on $\mathbf{V} \otimes \mathbf{W}$ by $g(v \otimes w) = gv \otimes gw$. It is only necessary to verify that this preserves the defining relations:

$$g(fv \otimes w) = fgv \otimes gw = gv \otimes fgw = gv \otimes g(fw) = g(v \otimes fw)$$

$$g((x + y) \otimes z) = (gx + gy) \otimes gz = gx \otimes gz + gy \otimes gz$$

$$= g(x \otimes z) + g(y \otimes z) = g(x \otimes z + y \otimes z)$$

$$g(z \otimes (x + y)) = gz \otimes (gx + gy) = gz \otimes gx + gz \otimes gy$$

$$= g(z \otimes x + z \otimes y)$$

This can be shown to give precisely the Kronecker product representation.

A major use of tensor products in group representation theory is the construction of new representations. In addition to the tensor product itself of two representations, we describe two other important methods.

First, we take the induced representation. Let H be a subgroup of G with $[G : H] = k$ and let M be an $\mathrm{F(H)}$-module of dimension n. Here $[G : H]$ is the index of H in G. Then

$$\mathrm{F(G)} \underset{\mathrm{F(H)}}{\otimes} M$$

is a representation of G of dimension kn called the ***induced representation.*** For groups of prime power order and some other types of groups, every irreducible representation is induced from a 1-dimensional representation of a subgroup. Richard Brauer proved that for any finite group, representations induced from 1-dimensional representations give a set of additive generators for the character ring.

Another method is especially important for representations of compact continuous groups. Let $\mathbf{V}$ be a complex representation of any group G. Then $\mathbf{V^n} = \mathbf{V} \otimes \mathbf{V} \otimes \ldots \otimes \mathbf{V}$ is a G-module. However, it is also a module over the symmetric group acting by permutation of the factors: $\pi(x_1 \otimes x_2 \otimes \ldots \otimes x_n) = x_{(1)\pi} \otimes x_{(2)\pi} \otimes \ldots \otimes x_{(n)\pi}$. Moreover these two actions commute: $g\pi(x_1 \otimes x_2 \otimes$

$\ldots \otimes x_n) = gx_{(1)\pi} \otimes gx_{(2)\pi} \otimes \ldots \otimes gx_{(n)\pi} = \pi(gx_1 \otimes gx_2 \otimes \ldots \otimes gx_n) = \pi g(x_1 \otimes x_2 \otimes \ldots \otimes x_n)$. The G-module $\mathbf{V}^{\mathbf{n}}$ decomposes as a direct sum according to the different irreducible representations of the symmetric group S_n.

One way to see this is to recall that corresponding to each irreducible representation there will be a central idempotent e_i in $\mathsf{C}(S_n)$, and any $\mathsf{C}(S_n)$-module M is the direct sum of $e_i M$. Since e_i is a sum of permutations π it will commute with any element G. Therefore $g(e_i M) = e_i g M \subset e_i M$ so $e_i M$ is a submodule over G.

EXAMPLE 5.5.4. For $n = 2$, let $S_2 = \{e, t\}$ where $t^2 = 2$. Then the two central idempotents are $e_1 = \frac{1}{2}(e + t)$ and $e_2 = \frac{1}{2}(e - t)$. Any S_2-module M splits as a direct sum of $e_1 M$ and $e_2 M$ since $e_1 e_2 = 0$ and $e_1 + e_2 = 1$.

Then $M \otimes M$ is acted upon by S_2 by $e(x \otimes y) = x \otimes y$ and $t(x \otimes y) = y \otimes x$. The submodule $e_1 M$ is spanned by $x \otimes y + y \otimes x$. On it S_2 acts trivially, that is, every operation is the identity. It is called the ***symmetric 2nd power*** of M.

The submodule $e_2 M$ is spanned by $x \otimes y - y \otimes x$. On it S_2 acts in such a way that interchanging the factors reverses the sign of any element. It is called the ***exterior 2nd power*** of M.

In general we can get many representations of any group from the tensor powers M^n of any 1–1 representation. The exterior powers are especially important.

DEFINITION 5.5.3. Let M be a $\mathsf{C}(\mathsf{G})$-module. The *kth exterior power* $\lambda^k(M)$ is given by $e_k M^k$ where e_k is the central idempotent

$$\frac{1}{n!} \Sigma\, \sigma(\pi)\pi$$

where $\sigma(\pi)$ is the sign of π. The *kth symmetric power* $s_k(M)$ is given by $f_k M^k$ where f_k is the idempotent

$$\frac{1}{n!} \Sigma\, \pi$$

EXAMPLE 5.5.5. For $k = 1$, $s_1(M)$ is M. For $k = 2$, $\lambda^2(M)$, $s_2(M)$ are in the preceding example.

DEFINITION 5.5.4. $x_1 \wedge x_2 \wedge \ldots \wedge x_k = e_k(x_1 \otimes x_2 \otimes \ldots \otimes x_k)$, where e_k is the central idempotent.

PROPOSITION 5.5.3. *Let $v_1, v_2, \ldots, v_n$ be a basis for M. Then a basis for $\lambda^k(M)$ is given by $v_{i_1} \wedge v_{i_2} \wedge \ldots \wedge v_{i_k}$ such that $i_1 < i_2 < \ldots < i_k$.*

Proof. We have from the definition $\pi(x_1 \wedge x_2 \wedge \ldots \wedge x_k) = \sigma(\pi)\,(x_1 \wedge x_2 \wedge \ldots \wedge x_k)$. The $v_{i_1} \wedge v_{i_2} \wedge \ldots \wedge v_{i_k}$ for arbitrary $i_1, i_2, \ldots, i_k$ form a spanning set for $\lambda^k(M)$ since $\lambda^k(M) = e_i M^k$ and $v_{i_1} \otimes v_{i_2} \otimes \ldots \otimes v_{i_k}$ form a basis for the tensor product. By applying a permutation π we may assume $i_1 \leqslant i_2 \leqslant \ldots \leqslant i_k$, and π will at most alter the sign of the result. If $v_{i_r} = v_{i_s}$ then let $\pi \in S_n$ interchange r, s. Then $\pi(v_{i_1} \wedge v_{i_2} \wedge \ldots \wedge v_{i_k}) = v_{i_1} \wedge v_{i_2} \wedge \ldots \wedge v_{i_k}$ since $i_r = i_s$ but $\pi(v_{i_1} \wedge v_{i_2} \wedge \ldots \wedge v_{i_k}) = \sigma(\pi)(v_{i_1} \wedge v_{i_2} \wedge \ldots \wedge v_{i_k}) = -(v_{i_1} \wedge v_{i_2} \wedge \ldots \wedge v_{i_k})$. So $v_{i_1} \wedge v_{i_2} \wedge \ldots \wedge v_{i_k} = -(v_{i_1} \wedge v_{i_2} \wedge \ldots \wedge v_{i_k})$. So it is zero. Therefore for $i_1 < i_2 < \ldots < i_k$ we have a spanning set.

The given terms are independent in M^k since different ones involve distinct basis elements $v_{i_1} \otimes v_{i_2} \otimes \ldots \otimes v_{i_k}$. □

EXAMPLE 5.5.6. A basis for $\lambda^2(M)$ is given by $v_i \wedge v_j$ where $i < j$. We have $v_j \wedge v_i = -v_i \wedge v_j$.

PROPOSITION 5.5.4. *Let a matrix have eigenvalues $e_1, e_2, \ldots, e_n$. Then the linear transformation it induces on the kth exterior power has eigenvalues all products $e_{i_1} e_{i_2} \ldots e_{i_k}$, $i_1 < i_2 < \ldots < i_k$.*

Proof. Choose a basis $v_1, v_2, \ldots, v_k$ in which M is lower triangular with main diagonal elements $e_1, e_2, \ldots, e_n$. Then $Mv_i = e_i v_i$ plus a linear combination of $v_j, j < i$. On the exterior power it will send $v_{i_1} \wedge v_{i_2} \wedge \ldots \wedge v_{i_k}$ to $e_{i_1} e_{i_2} \ldots e_{i_k}(v_{i_1} \wedge v_{i_2} \wedge \ldots \wedge v_{i_k})$ to plus terms involving lower v_j. So it will be lower triangular with eigenvalues $e_{i_1} e_{i_2} \ldots e_{i_k}$. □

We next apply these results to determine the character ring of U_n, the $n \times n$ unitary matrices, under some assumptions. Let D_n^1 the diagonal matrices such that $|d_{ii}| = 1$ for $i = 1$ to n. Then D_n^1 is commutative. Every element of U_n is similar to a matrix of D_n^1, hence the characters take the same value. Therefore any character is determined by its values on D_n^1. The group D_n^1 is ξ_1^n where ξ_1 is the group of complex numbers of absolute value 1.

We have shown in an earlier exercise that every irreducible representation of a finite commutative group is 1-dimensional. This result extends to compact commutative groups (in effect we can simultaneously diagonalize the matrices). Every 1-dimensional representation is equivalent to a unitary representation, which is a homomorphism $\xi_1^n \to \xi_1$. Such representation are sums of homomorphisms $\xi_1 \to \xi_1$, which are of the form $x \to x^n$. Let x_i denote the homomorphism $\xi_1^n \to \xi$ which takes $(a_1, a_2, \ldots, a_n)$ to a_i. Then all 1-dimensional representations of D_n^1 are of the form $x_1^{m_1} x_2^{m_2} \ldots x_n^{m_n}$ for $m_i \in \mathbf{Z}$.

Therefore any character of U_n can be described by a finite sum

$$\sum_{m_1, \ldots, m_n \in \mathbf{Z}} x_1^{m_1} x_2^{m_2} \ldots x_n^{m_n} p(m_1, m_2, \ldots, m_n)$$

where $p(m_1, m_2, \ldots, m_n)$ is a positive integer.

EXAMPLE 5.5.7. The standard representation $r_I : \mathcal{U}_n \to \mathcal{U}_n$ restricted to D_n^1 is a direct sum of the n representations $x_1, x_2, \ldots, x_n$. So its character is $x_1 + x_2 + \ldots + x_n$.

EXAMPLE 5.5.8. The determinant d gives a representation $\mathcal{U}_n \to \mathcal{U}_1$. Its value on D_n^1 is the product of the main diagonal entries. So it has character $x_1 x_2 \ldots x_n$. Its complex conjugate has character $(x_1 x_2 \ldots x_n)^{-1}$.

EXAMPLE 5.5.9. The kth exterior power $\lambda^k(M)$ has character $\sum_{i_1 < \ldots < i_k} x_{i_1} x_{i_2} \ldots x_{i_k}$ by the Proposition 5.5.4.

The character of any finite dimensional representation r of $\mathcal{U}_n$ must be a function of $x_1, x_2, \ldots, x_n$ which is symmetric under interchanging the x_i. The reason is that if P is the matrix of a permutation interchanging i, j then PAP^{-1} for $A \in D_n^1$ has the effect of interchanging the ith and jth main diagonal entries. Therefore it interchanges x_i, x_j. This will not change the characters since $P \in \mathcal{U}_n$ and $\mathrm{Tr}(r(P)r(A)r(P)^{-1}) = \mathrm{Tr}(r(A))$.

DEFINITION 5.5.5. The *kth elementary symmetric function* σ_k of variables $x_1, x_2, \ldots, x_n$ is $\sum_{i_1 < \ldots < i_k} x_{i_1} x_{i_2} \ldots x_{i_k}$.

EXAMPLE 5.5.10. For $n = 4$ these are $\sigma_1 = x_1 + x_2 + x_3 + x_4$, $\sigma_2 = x_1 x_2 + x_1 x_3 + x_1 x_4 + x_2 x_3 + x_2 x_4 + x_3 x_4$, $\sigma_3 = x_1 x_2 x_3 + x_1 x_2 x_4 + x_1 x_3 x_4 + x_2 x_3 x_4$, $\sigma_4 = x_1 x_2 x_3 x_4$.

THEOREM 5.5.5. *Let $\mathcal{R}$ be a commutative ring with unit. Then any symmetric polynomial in $x_1, x_2, \ldots, x_n$ with coefficients in $\mathcal{R}$ is a polynomial in $\sigma_1, \sigma_2, \ldots, \sigma_n$ with coefficients in $\mathcal{R}$.*

Proof. Let $p(x_1, x_2, \ldots, x_n)$ be any symmetric polynomial. For any product Γ of powers of the x_i let $\Sigma\,[\Gamma]$ denote the sum of all terms symmetric to Γ. It suffices to show all the $\Sigma\,[\Gamma]$ can be expressed as integer polynomials in $\sigma_1, \sigma_2, \ldots, \sigma_n$ since every polynomial $p(x)$ is a sum of expressions $\Sigma\,[\Gamma]$ with coefficients in $\mathcal{R}$. If $n = 1$ or the degree of $p(x) = 1$ the theorem is true.

Assume that this result is true for $m < n$ and for $m = n$ and expression $\Sigma\,[\Gamma]$ of degree less than the degree of $p(x)$. If Γ involves every variable, $\Sigma\,[\Gamma]$ is divisible by $\sigma_n = x_1 x_2 \ldots x_n$. By induction we can express

$$\frac{\Sigma\,[\Gamma]}{\sigma_n}$$

as a polynomial in the σ_i. So $\Sigma\,[\Gamma]$ is such a polynomial.

Suppose Γ does not involve every variable, but omits variable x_n. Let $\Sigma\, s[\Gamma]$ denote the sum of all terms involving $x_1, x_2, \ldots, x_{n-1}$ symmetric to Γ. Let v_i be the elementary symmetric functions in $x_1, x_2, \ldots, x_{n-1}$. We have an expression by induction $\Sigma\, s[\Gamma] = f(v_1, v_2, \ldots, v_{n-1})$. Let $g = \Sigma\, [\Gamma] - f(\sigma_1, \sigma_2, \ldots, \sigma_{n-1})$. If we substitute zero for x_n we obtain $\Sigma\, s[\Gamma] - f(v_1, v_2, \ldots, v_{n-1}) = 0$. By symmetry the same is true if we substitute zero for any x_i. This means all terms of g involve every x_i. Therefore g is divisible by σ_n and we can express it in terms of $\sigma_1, \sigma_2, \ldots, \sigma_n$ as before. □

EXAMPLE 5.5.11. To express $p(x) = x_1^2 x_2 + x_1^2 x_3 + x_2^2 x_1 + x_2^2 x_3 + x_3^2 x_1 + x_3^2 x_2$ in $\sigma_1, \sigma_2, \sigma_3$, we first consider $x_1^2 x_2 + x_2^2 x_1$. This is $(x_1 + x_2) x_1 x_2 = v_1 v_2$. Then take $p(x) - \sigma_1 \sigma_2 = 3 x_1 x_2 x_3 = 3\sigma_3$. Therefore $p(x) = \sigma_1 \sigma_2 + 3\sigma_3$.

DEFINITION 5.5.6. The *character ring* of a group G is the ring of complex functions $G \to \mathbf{C}$ generated by the characters.

EXAMPLE 5.5.12. For $G = \mathbf{Z}_2$, the character ring is the ring of all sums $n(1, 1) + m(1, -1)$ since $(1, 1)$ and $(1, -1)$ are the irreducible representations.

THEOREM 5.5.6. *The character ring of U_n consists of all functions $\sigma_n^k f(\sigma_1, \sigma_2, \ldots, \sigma_n)$ where k is an integer and f is an integer valued polynomial.*

Proof. We have a function symmetric in $x_1, x_2, \ldots, x_n$. Write it as $(x_1 x_2 \ldots x_n)^{-k} f(x_1, x_2, \ldots, x_n)$ where f involves only positive powers of each x_i, by taking k large. Then f is a symmetric polynomial in $x_1, x_2, \ldots, x_n$. By the last theorem, it has the given form.

Since $\lambda^i r_I$ has character σ_i, we can obtain any polynomial $f(\sigma_1, \sigma_2, \ldots, \sigma_n)$ as a sum or difference of characters of actual representatives. And σ_n^{-1} is the character of the conjugate of the determinant representation. Therefore we can obtain all the functions described. □

It is important to note that the character ring contains negatives and differences of representations, so all its elements are not actually characters of a representation. A study of irreducible representations of U_n requires a study of irreducible representations of S_n, and is more complicated.

EXERCISES

In the following exercises ξ_1 is the multiplicative group of complex numbers e^{ix} of absolute value 1 and D_n^1 is the group of $n \times n$ complex diagonal matrices whose main diagonal entries have absolute value 1. Other notation is the same as in the preceding text.

Level 1

1. Express $x_1^3 x_2 + x_1^3 x_3 + x_2^3 x_1 + x_2^3 x_3 + x_3^3 x_1 + x_3^3 x_2$ as a polynomial in $\sigma_1, \sigma_2, \sigma_3$. First study $x_1^3 x_2 + x_2^3 x_1$.

2. Tell why every element of the character ring of a compact group is a sum $\Sigma\, n_i \chi_i$ where $n_i \in \mathbf{Z}$ and χ_i are characters of irreducible representations. Show every member is a difference of the characters of two actual representations.
3. The irreducible representations of ξ_1 have the form $e^{2\pi i\theta n}$, $n \in \mathbf{Z}$ where θ ranges from 0 to 2π. The orthogonality relations for the characters have the form

$$\int_0^{2\pi} e^{2\pi i\theta n}\, e^{-2\pi i\theta m}\, d\theta = 0$$

for $m \neq n$. Verify these by integration.
4. Let M be the direct sum of m copies of R and let N be the direct sum of n copies of R. What is $M \otimes N$?
5. Prove $\mathbf{Z_m} \otimes \mathbf{Z_n} = \{0\}$ if m and n are relatively prime. The tensor product is taken over $\mathbf{Z}$.
6. Prove $\mathbf{Z_n} \otimes \mathbf{Z_n} = \mathbf{Z_n}$.
7. What is $\mathbf{Z_n} \otimes \mathbf{Q}$?
8. Prove the regular representation of a group is induced from a trivial representation of the subgroup $\{e\}$.
9. Show the group ring $C(G \times H)$ is isomorphic to $C(G) \underset{\mathbf{C}}{\otimes} C(H)$.

Level 2

1. What is the dimension of $\lambda^i(r_I)$?
2. What is the character of $s_i(r_I)$?
3. Corresponding to the three irreducible representations of S_3 we have $M \otimes M \otimes M \simeq \lambda^2(M) + s_2(M) + M_3$ for some M_3. By subtraction find the dimension of M_3 if M has dimension n. For the unitary group, find its character.
4. The tensor product can be defined more precisely as follows. Let G be the set of all functions f: $M \times N \to \mathbf{Z}$, which are zero on all but a finite set of ordered pairs. Let $H(g)$ be the subgroup generated by all functions of the form $g(xr, z) - g(x, rz)$, $g(x + y, z) - g(x, z) - g(y, z)$, $g(x, z + w) - g(x, z) - g(w, z)$. Then the tensor product is the quotient $G/H(g)$. Let h be the mapping $M \times N \to H(g)$ such that $h(m, n)$ is 1 on (m, n) and 0 on all other pairs. Prove h is bilinear.
5. Let G' be an arbitrary abelian group and b any bilinear mapping $M \times N \to G'$. Define $r : G \to G'$ by $r(f) = \Sigma\, b(m, n) f(m, n)$. Prove r is 0 on $H(g)$ and $rh = b$ where $M, N, G, H(g), h$ are the same as in Exercise 4.
6. Prove r is unique in the above exercise.
7. Show that the irreducible 2-dimensional representation of S_3 is induced from a representation of $\mathbf{Z_3}$.
8. Find a conjugation by a permutation matrix interchanging the first two factors of D_n^1.

Level 3

1. Find a relation between λ^n and the determinant of an $n \times n$ matrix.
2. Prove a complex representation cannot be decomposed as a direct sum if there does not exist any matrix other than the identity which commutes with every matrix of the representation.
3. Prove the representation $\lambda^i r_I$ of the unitary group is irreducible (How could its character be a sum of positive symmetric polynomials in the x_i?)
4. Try to find a relation between characters of ξ_1 and Fourier series (consider Exercise 3, Level 1).
5. Prove from the definition that $(M \otimes N) \otimes T \simeq (M \otimes T) \otimes (N \otimes T)$ where M, N, and T are modules.
6. Describe an induced representation as a matrix.
7. Prove that tensor products of modules correspond to Kronecker products of representations.
8. Find the character ring of unitary matrices of determinant 1.
9. Prove the ring $M_n(\mathbf{C})$ is a semigroup ring (of matrices having at most one 1 entry) but not a group ring for n prime. Use the fact that a group of order n^2 is commutative if n is prime.

CHAPTER 6

Field theory

A large part of field theory is concerned with systems of numbers intermediate between **Q** and **C** but closed under addition, multiplication and division. Examples are $\mathbf{Q}(\sqrt{5})$, the set of all numbers $a + b\sqrt{5}$ where $a, b \in \mathbf{Q}$, $\mathbf{Q}(\sqrt[3]{3})$, the set of all numbers $a + b\sqrt[3]{3} + c\sqrt[3]{9}$. These examples are finite dimensional over **Q**. Fields of this nature are involved in solution of algebraic equations. They can be defined in terms of quotients of a polynomial ring $\mathbf{F}[x]$ by the ideal $p(x)\mathbf{F}(x)$ where $p(x)$ is an irreducible polynomial such that $p(y)$ is zero for a generator y of the field.

Ruler and compass constructions involve fields obtained by repeatedly adding square roots to the rationals. This gives a sequence of degree 2 extensions. Therefore a number such as $\sqrt[3]{2}$ which generates a degree 3 extension cannot be obtained by ruler and compass constructions. As a consequence various problems such as trisecting an angle and duplicating a cube by ruler and compass are impossible. ***Galois theory*** associates a group to every field extension which is generated by a complete set of roots of a polynomial. It is a group of permutations of the x_i which give field automorphisms, like $i \to -i$. A polynomial is solvable by radicals if and only if its group is solvable. It can be proved in this way that there exist polynomials of the 5th degree not solvable by extraction of square, cube, fourth, and fifth roots.

There exists a unique finite field of any prime power order p^n. It can be obtained by an extension generated by a root of any irreducible degree n polynomial over $\mathbf{Z_p}$.

A ***t-error correcting code*** is a set C of sequences of symbols S or words such that if at most t errors are made in a sequence and at most t errors are made in another sequence the result will be different. Therefore if at most t errors are made the receiver can uniquely determine the correct original word, despite the errors. These are used in computers and have other communications applications. A code is called ***perfect*** if it requires a minimal number of digits added to the original code according to a theoretical inequality. All perfect codes have been determined. The most important infinite family are called the ***Hamming codes.***

Other non-perfect but frequently effective error-correcting codes have as the set of code words an ideal in the ring of polynomials subject to a relation $x^n = 1$. They are the cyclic codes. These include BCH, Reed–Solomon, and Fire codes. The latter are designed to correct a burst of errors occurring in a short sequence within the message block.

A ***Latin square*** is an $n \times n$ table in which each entry occurs exactly once in every row and column. These can be used to arrange statistical testing of several controlled factors for convenient analysis. For four or more factors orthogonal Latin squares are used. In these every possible pair of entries can be obtained as a pair of entries in the same location in two Latin squares. We show how to construct orthogonal Latin squares for all orders not of the form $4n + 2$, from finite rings.

A ***projective plane*** is a system of any objects called ***lines*** and ***points***, in which every two points determine a unique line, and every two lines have a unique point in common. These can be constructed from finite fields for all prime power orders, but for other orders it is not known whether they exist. A projective plane is one kind of ***balanced incomplete block design***. Such designs are also used in statistics.

6.1 FINITE DIMENSIONAL EXTENSIONS

A large part of field theory was motivated by the problem of solving polynomial equations $x^n + c_1 x^{n-1} + \ldots + c_n = 0$ and studying of the numbers, such as $2 + 3\sqrt{5}$, which occur as solutions. For a given polynomial, the solutions generate a field containing the rationals. Unlike the real or complex numbers, the dimension of this field over the rationals will be finite. Such a field is called a ***finite dimensional extension*** of the rationals.

Finite fields are necessarily finite extensions of a field $\mathbf{Z_p}$.

DEFINITION 6.1.1. A field $\mathbf{E}$ is an *extension* of $\mathbf{F}$ if $\mathbf{F} \subset \mathbf{E}$. The *degree* $[\mathbf{E}:\mathbf{F}]$ of the extension is the dimension of $\mathbf{E}$ as a vector space $\mathbf{V}$ over $\mathbf{F}$. $\mathbf{F}$ is called a *subfield* of $\mathbf{E}$.

EXAMPLE 6.1.2. $[\mathbf{C}:\mathbf{R}] = 2$.

If $[\mathbf{E}:\mathbf{F}]$ is finite then every element of $\mathbf{E}$ is algebraic over $\mathbf{F}$, that is, satisfies a polynomial with coefficients in $\mathbf{F}$. The reason is, the powers $1, x, x^2, \ldots, x^n$ lie in an n-dimensional space, if $[\mathbf{E}:\mathbf{F}] = n$, and are therefore linearly dependent. The polynomial of lowest degree satisfied by x (with first coefficient 1) is unique (else take the difference of two) and cannot be factored over $\mathbf{F}$, since nonzero elements have inverses. It is called the ***minimum polynomial*** of x.

EXAMPLE 6.1.2. The minimum polynomial of $\sqrt{2}$ over the rationals is $x^2 - 2 = 0$.

If $\mathbf{E} \supset \mathbf{F}$ and S is a subset of $\mathbf{E}$ the set of members of $\mathbf{E}$ obtained by adding, subtracting, multiplying, and dividing finitely many times elements of $S \cup \mathbf{F}$ is itself a field, denoted $\mathbf{F}(S)$. It is called the ***field generated*** by S over $\mathbf{F}$.

EXAMPLE 6.1.3. The number $\sqrt{3}$ generates the field $\{a + b\sqrt{3},\ a, b \in \mathbf{Q}\}$ over $\mathbf{Q}$.

DEFINITION 6.1.2. A polynomial is *monic* if its first coefficient is 1. It is *irreducible* if it cannot be factored as a product of two polynomials of lower degree.

THEOREM 6.1.1. *If the minimum polynomial of x over* $\mathbf{E}$ *has degree n then $1, x, x^2, \ldots, x^{n-1}$ form a basis for* $\mathbf{E}(x)$ *over* $\mathbf{E}$. *The value of a polynomial $p(t)$ for $t = x$ is zero if and only if $p(t)$ is divisible by the minimum polynomial of x.*

Proof. If $1, x, x^2, \ldots, x^{n-1}$ were linearly dependent x would satisfy a polynomial of lower degree. Let R be the set of all expressions $a_1 + a_2 x + a_3 x^2 + \ldots + a_n x^{n-1}$, $a_i \in \mathbf{F}$. Then R is closed under addition and multiplication by elements of $\mathbf{F}$. Let the minimum polynomial of x be written as $x^n = -c_1 x^{n-1} - \ldots - c_{n-1} x - c_n$. This expresses x^n as an element of R. Suppose $x^k \in R$. Then x^k is a linear combination of $1, x, x^2, \ldots, x^{n-1}$. So x^{k+1} is a linear combination of $x, x^2, \ldots, x^n$ but x^n is a linear combination of $1, x, x^2, \ldots, x^{n-1}$. Therefore by induction R contains all x^k and is therefore closed under products.

Let $p(x) = 0$ and let $f(x)$ be the minimum polynomial of x. Then $p(t) = g(t)f(t) + r(t)$ where $\deg r(t) < \deg f(t)$. So $r(t) = 0$. So $f(t)$ divides $p(t)$.

The ring R has no zero divisors since it is contained in a field. So cancellation holds. For any $z \in R$, $z \neq 0$, the mapping $z \to zy$ is 1-1 and linear. Since it has kernel zero its image must have dimension n. So $zy = 1$ for some y. So z has a multiplicative inverse. So R is a field. It is contained in $\mathbf{F}(x)$. Since it is closed under addition, subtraction, multiplication, and division, and contains x, it must equal $\mathbf{F}(x)$. □

There is a similar result which can be used to construct extension fields as quotients of polynomial rings.

THEOREM 6.1.2. *Let $g(x)$ be an irreducible monic polynomial over* $\mathbf{F}$, *and let* $\mathbf{F}[x]$ *denote the ring of polynomials in x with coefficients in* $\mathbf{F}$. *Then*

$$\frac{\mathbf{F}[x]}{g(x)\mathbf{F}[x]}$$

is a field $\mathbf{E}$ *with basis $1, x, x^2, \ldots, x^{n-1}$ over* $\mathbf{F}$ *and the minimum polynomial of x in* $\mathbf{E}$ *is $g(x)$. If y is any element of any extension field satisfying $g(y) = 0$ then there exists an isomorphism* $\mathbf{E} \to \mathbf{F}(y)$ *sending x to y.*

Proof. Let
$$\mathbf{E} = \frac{\mathbf{F}[x]}{g(x)\mathbf{F}[x]}.$$

Then $\mathbf{E}$ is a ring containing $\mathbf{F}$. Any polynomial $h(x)$ is equivalent in $\mathbf{E}$ to a polynomial of degree $< n = \deg(g(x))$, since $h(x) = g(x)g(x) + r(x)$ so $h(x) - r(x) \in g(x)\mathbf{F}(x)$. Let $h(x)$ be a nonzero polynomial of degree $< n$. Since $g(x)$ is irreducible (prime) there exist $r(x)$, $s(x)$ such that $r(x)h(x) + g(x)s(x) = 1$. So $r(x)h(x)$ is equivalent to 1 in $\mathbf{E}$. Therefore $h(x)$ has a multiplicative inverse. Therefore $\mathbf{E}$ is a field.

Since any polynomial is equivalent to a polynomial of degree less than n, the powers $1, x, x^2, \ldots, x^{n-1}$ are a spanning set. No polynomial of degree less than n is divisible by $g(x)$. Therefore they are independent.

Since $g(x) = 0$ in $\mathbf{E}$ but no polynomial of lower degree is zero in $\mathbf{E}$ unless it is zero in $\mathbf{F}[x]$, $g(x)$ is the minimum polynomial of x.

Let y be an element of $\mathbf{E}_1 \supset \mathbf{F}$ such that $g(y) = 0$. There exists a map $\mathbf{F}[x]$ into $\mathbf{F}(y)$ sending any polynomial $p(x)$ to $p(y)$. This map is a ring homomorphism. By the previous theorem it is onto, since $1, y, y^2, \ldots, y^{n-1}$ form a basis. It sends $g(x)$ to $g(y) = 0$. So $g(x)\mathbf{F}[x]$ is in its kernel. This gives a mapping from
$$\mathbf{E} = \frac{\mathbf{F}[x]}{g(x)\mathbf{F}[x]}$$
onto $\mathbf{F}(y)$. It is onto. Since both spaces have dimension n, it is an isomorphism. □

This result gives a method of constructing many finite dimensional extensions of $\mathbf{F}$. Take any irreducible polynomial $p(x)$. Then
$$\frac{\mathbf{F}[x]}{p(x)\mathbf{F}[x]}$$
is a finite dimensional extension.

Moreover every finite dimensional extension $\mathbf{E}_1$ can be obtained by repeating this procedure. Let $y \in \mathbf{E}_1$ but $y \notin \mathbf{E}$. Then the subfield generated by y is isomorphic to a field
$$\frac{\mathbf{F}[x]}{p(x)\mathbf{F}[x]}$$
Then an element of $\mathbf{E}_1 \backslash \mathbf{F}(y)$ can be taken, and so on.

To carry this out, some criteria for irreducibility of a polynomial are needed. The following are some well-known facts:

Ir-1. If a polynomial $f(x)$ with integer coefficients factors over $\mathbf{Q}$, it factors into two polynomials with integer coefficients. The first and last coefficients of the factors divide those of $f(x)$.

Ir-2. (Eisenstein Irreducibility Criterion.) Suppose that $f(x+k)$ for some integer k has the property that f is monic of degree n, some prime p divides the coefficients of $x^{n-1}, n^{n-2}, \ldots, x, 1$ and p^2 does not divide the constant term. Then f is irreducible.

Ir-3. A polynomial of degree 2, or 3 is irreducible if it has no roots in **F**.

EXAMPLE 6.1.4. The polynomial x^3+2x^2+2x+2 is irreducible by the second criterion, since 2 divides the coefficients of $x^2, x, 1$ but 2^2 does not divide the constant. So we have an extension field. A basis is given by $1, x, x^2$. Products can be computed from the table:

	1	x	x^2
1	1	x	x^2
x	x	x^2	$-2x^2-2x-2$
x^2	x^2	$-2x^2-2x-2$	$2x^2+2x+4$

Inverses can be computed from the Euclidean algorithm, finding $r(x), s(x)$ such that $r(x)p(x)+s(x)f(x)=1$.

For instance, if we want to find the inverse of x^2+x+1, divide it into x^3+2x^2+2x+2. The quotient is $x+1$ and the remainder is 1. Therefore $(x^2+x+1)(x+1)-(x^3+2x^2+2x+2)=1$. So $x+1$ is $(x^2+x+1)^{-1}$ in this field.

EXERCISES

Level 1

1. Find the minimum polynomial of $\sqrt{2}+\sqrt{3}$. (Compute its first four powers and show they are linearly dependent.)
2. Prove the 3rd irreducibility criterion mentioned.
3. Show the polynomial x^2+1 is irreducible over **R**. Compute the multiplication table for $1, x$.
4. Show the extension in Exercise 2 is isomorphic to the complex numbers, using Theorem 6.1.2.
5. Show that two different irreducible quadratic polynomials can give the same extension field.

Level 2

1. Suppose x, y lie in extension fields of respective degree m, n over **F**. Show both lie in an extension field of degree at most mn over **F**. (Adjoin first x, then y.)
2. Show that x^3+3 is irreducible. (Any roots would have to be integers dividing 3.)

3. Write out the multiplication table of 1, x, x^2 in

$$\mathbf{E} = \frac{\mathbf{Q}[x]}{(x^3+3)\mathbf{Q}[x]}$$

4. Multiply $(1 + x + x^2)(2 + 3x - x^2)$ in this extension field.
5. Find an inverse of $(1 + x)$ in **E**.
6. Verify that Exercise 5 gives an inverse of $1 + \sqrt[3]{3}$.

Level 3

1. There is a finite procedure for determining if a polynomial $p(x)$ of degree 2, 3 is irreducible over **Z**. (Try linear factors over **Z** whose first and last coefficients divide those of $p(x)$.) Is there such a procedure for higher degrees, where some factors may not be linear? Can the coefficients of the factors be bounded (consider the roots of $p(x)$)?
2. Prove that **C** is the only finite extension field of **R**. Use the fact that every polynomial of odd degree has a real root.
3. Prove that there exist infinitely many non-isomorphic extensions of **Q** of degree 2.
4. Show that $(x^3 - 3)$ has a root in $\mathbf{Q}(\sqrt[3]{3})$ but does not factor into linear factors over this field.
5. Find an irreducible polynomial of degree 3 over **Q** which does factor into linear factors of a single root is adjoined to **Q**. (Look up cubic equations, the discriminant must be a perfect square.)

6.2 APPLICATIONS OF EXTENSIONS OF THE RATIONALS

Field theory can be used to study solvability of polynomial equations under various restrictions. This in turn gives conditions for solving problems depending on polynomial equations.

An example is ruler and compass constructions. Without being formal, a unit length is given. Then it is asked whether, using an ideal (unmarked ruler) and an ideal compass a given geometrical figure can be produced. For example regular triangles, squares, and pentagons can be drawn by these constructions. Lines may be bisected and perpendiculars drawn at any point.

Choose coordinates such that the given segment has endpoints (0, 0) and (1, 1).

THEOREM 6.2.1. *A point can be constructed by ruler and compass if and only if its coordinates can be obtained from the rational numbers by repeatedly extracting square roots.*

Proof. There exist constructions for adding, subtracting, multiplying, dividing, and taking square roots. Thus coordinates of this form can be separately generated as lengths. Then by laying out these lengths on the coordinate axes we may obtain the required point.

Conversely the only way we can obtain new points is by intersecting two lines (ruler), a line and a circle (ruler and compass), or two circles (compass). The line must be between two points already constructed, so its coefficients lie in the field they generate. The circle must have a radius consisting of a length already constructed and a center already constructed, so its coefficients lie in the field corresponding to points and lengths already generated.

The intersection of two lines is found by solving simultaneous linear equations. Therefore it lies in the same field since such a solution can be obtained by adding, subtracting, multiplying, and dividing.

An intersection of a line and a circle

$$\begin{cases} ax + by = c \\ (x-h)^2 + (y-k)^2 = r^2 \end{cases}$$

can be found by solving the linear equation for one variable and substituting in the equation for the circle. This involves only a quadratic equation, which lies in an extension field obtained by adding a square root.

The intersection of two circles

$$\begin{cases} (x-a)^2 + (y-b)^2 = c^2 \\ (x-h)^2 + (y-k)^2 = r^2 \end{cases}$$

may be found in the same way after first subtracting one equation from the other to obtain a linear equation. □

COROLLARY 6.2.2. *The coordinates of a point constructible by ruler and compass lie in an extension* $\mathbf{E}$ *of the rationals* $\mathbf{Q}$ *such that there exists* $\mathbf{Q} = \mathbf{E}_1 \subset \mathbf{E}_2 \subset \ldots \subset \mathbf{E}_n = \mathbf{E}$ *and* $[\mathbf{E}_{i+1} : \mathbf{E}_i] = 2$ *for each* i.

Proof. An extension field obtained by adding a square root corresponds to an irreducible quadratic. So the extension has degree 2, if it is nontrivial, by Theorem 6.1.1. □

PROPOSITION 6.2.3. *Suppose* $\mathbf{E} \supset \mathbf{F} \supset \mathbf{K}$ *and* $[\mathbf{E} : \mathbf{K}]$ *is finite. Then* $[\mathbf{E} : \mathbf{K}] = [\mathbf{E} : \mathbf{F}]\,[\mathbf{F} : \mathbf{K}]$.

Proof. All the extensions involved must be finite. Let x_i be a basis for $\mathbf{F}$ over $\mathbf{K}$ and y_j a basis for $\mathbf{E}$ over $\mathbf{F}$. Let $z \in \mathbf{E}$ then $z = \Sigma\, a_j y_j = \Sigma\, \Sigma\, b_{ji} x_i y_j$ for some $a_j \in \mathbf{F}$, $b_{ji} \in \mathbf{K}$. Therefore the $x_i y_j$ span $\mathbf{E}$ over $\mathbf{K}$. Suppose they are dependent. Let $\Sigma\, b_{ji} x_i y_j = 0$. Let $a_j = \Sigma\, b_{ji} x_i$. If some $b_{ij} \neq 0$ then some $a_j \neq 0$. Therefore $\Sigma\, a_j y_j \neq 0$. This is false. So $x_i y_j$ form a basis for $\mathbf{E}$ over $\mathbf{K}$. □

COROLLARY 6.2.4. *The coordinates of a point constructible by ruler and compass lie in an extension of the rationals of degree* 2^n.

COROLLARY 6.2.5. *If* x *is a number which satisfies an irreducible polynomial over the rationals of degree not a power of two, then* x *cannot be a length or coordinate constructible by ruler and compass.*

Proof. If x were constructible, $\mathbf{Q} \subset \mathbf{Q}(x) \subset \mathbf{E}$ where $[\mathbf{E}:\mathbf{Q}] = 2^n$. So $[\mathbf{Q}(x):\mathbf{Q}] | 2^n$. Therefore $[\mathbf{Q}(x):\mathbf{Q}] = 2^m$ for some m. So the minimum polynomial of x has degree 2^m. □

This criterion suffices for most of the impossibility results for ruler and compass construction. Trisection of an angle and duplication of a cube (construction of a cube with twice the volume of a given cube) require solving an irreducible cubic. Constructing a regular p-gon for p a prime requires finding a solution to an irreducible equation of degree $p - 1$ so p must have the form $2^n + 1$, as $2, 3, 5, 17, 257, \ldots$. Since π does not satisfy any polynomial over the rationals it cannot be constructed, so that a circle cannot be squared.

We give a brief survey of a topic which would require one or more chapters to deal with adequately, ***Galois theory***. There is a close relationship between solving a polynomial and studying how quantities are affected by permutations of the roots. For instance, the coefficients

$$\begin{aligned} -c_1 &= x_1 + x_2 + \ldots + x_n \\ c_2 &= x_1x_2 + x_2x_3 + \ldots + x_{n-1}x_n \\ &\quad \ldots\ldots\ldots \\ (-1)^n c_n &= x_1x_2 \ldots x_n \end{aligned}$$

of $x^n + c_1x^{n-1} + \ldots + c_n = (x - x_1)(x - x_2) \ldots (x - x_n)$ are unchanged by permutations of the x_i. By Theorem 5.5.5, any function of the x_i unchanged under permutations of the roots can be expressed as a rational function of $c_1, c_2, \ldots, c_n$. Therefore it can be computed. The discriminant function

$$\Delta = \prod_{i<j} (x_i - x_j) = (x_1 - x_2)(x_1 - x_3) \ldots (x_{n-1} - x_n)$$

at most changes sign under permutation of the roots. Therefore Δ^2 is symmetric and Δ can be computed from Δ^2 by extracting a square root. Equations of degrees 2, 3, 4 can be solved by first computing Δ and then a series of quantities less and less invariant under permutations, culminating in $x_1, x_2, \ldots, x_n$, the roots, invariant under no permutations. This is the reason that the roots of a quadratic appear as

$$\frac{-c_1 \pm \Delta}{2}$$

The $\pm$ corresponds to permuting the two roots.

This method corresponds with finding a series of fields $\mathbf{Q} = \mathbf{F_1} \subset \mathbf{F_2} \subset \ldots \subset \mathbf{F_n} = \mathbf{Q}(x_1, x_2, \ldots, x_n)$, where each $\mathbf{F_i}$ is obtained from $\mathbf{F_{i-1}}$ by extracting an nth root.

DEFINITION 6.2.1. The *Galois group* of an irreducible polynomial $p(x)$ over $\mathbf{F}$ is the group of all automorphisms α of the field $\mathbf{F}(x_1, x_2, \ldots, x_n)$ where the x_i are the roots of $p(x)$ such that $\alpha(a) = a$ for $a \in \mathbf{F}$.

EXAMPLE 6.2.1. The Galois group of an irreducible quadratic is $\mathbf{Z_2}$. The automorphism interchanges

$$x_1 = \frac{-c_1 + \Delta}{2}$$

with

$$x_2 = \frac{-c_1 - \Delta}{2}$$

It can be verified that Δ is irrational and, like complex conjugation, this process preserves sums, and products.

EXAMPLE 6.2.2. The Galois group over $\mathbf{Q}$ of $x^3 - 2$ is the symmetric group of all permutations of $\sqrt[3]{2}$, $\omega\sqrt[3]{2}$, $\omega^2\sqrt[3]{2}$ where ω is a non-real number such that $\omega^3 = 1$.

The Galois group of any polynomial is contained in the symmetric group of permutations of the x_i since if x satisfies $p(x) = 0$ its image under a field automorphism must also satisfy $p(x) = 0$.

Recall that a group G is solvable if and only if there exist subgroups $G = N_0 \supset N_1 \supset N_2 \supset \ldots \supset N_m = \{e\}$ where N_i is normal in N_{i-1} and G/N_1, $N_1/N_2, \ldots, N_{n-1}/N_m$, N_m are all commutative groups. Without loss of generality they may be taken as cyclic groups of prime order, if G is finite.

EXAMPLE 6.2.3. Any commutative group is solvable.

EXAMPLE 6.2.4. The symmetric groups of degree 2, 3, 4 are solvable.

It is proved in Galois theory that extracting a sequence of pth roots corresponding to a sequence of field extension $\mathbf{Q} = \mathbf{F_0} \subset \mathbf{F_1} \subset \mathbf{F_2} \subset \ldots \subset \mathbf{F_n}$ where the Galois group of $\mathbf{F_n}$ over $\mathbf{F_{k+1}}$ is a normal subgroup of that of $\mathbf{F_n}$ over $\mathbf{F_k}$ and the quotient group is $\mathbf{Z_p}$. Moreover the converse is also true. Therefore a polynomial is solvable by radicals if and only if its Galois group is solvable. There exists polynomials of degree 5 whose Galois group is the symmetric group. Since S_5 is not solvable, these polynomials cannot be solved by a series of processes involving extraction of nth roots.

EXERCISES

Level 1

1. Tell why this diagram, constructible by ruler and compass, multiplies lengths x, y, where AC $= x$, DE $= y$, BC $= 1$, AE $= xy$.

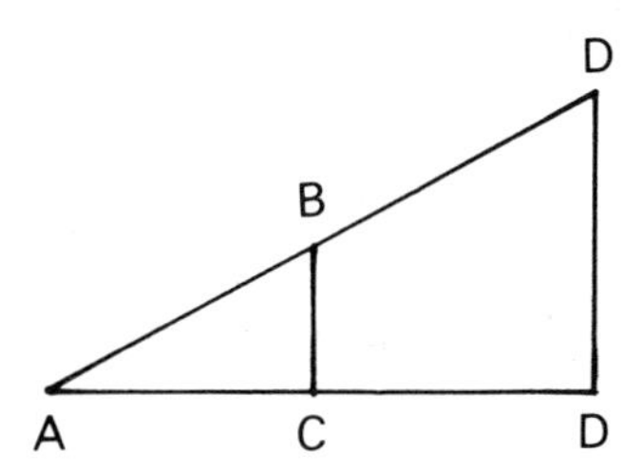

2. Relabel it to give a method of division.
3. Tell how to add or subtract two lengths by ruler and compass.
4. Duplication of a cube means given an edge x (set $x = 1$) of a cube construct the edge y of a cube having twice the volume. Show this leads to $y^3 - 2 = 0$. Prove this equation has no integer root. Therefore it is an irreducible cubic. Therefore duplication of the cube by ruler and compass is not possible.
5. The addition formulas for sine, cosine give $\cos 3\theta = 4\cos^3\theta - 3\cos\theta$. Let $3\theta = 60°$, that is we want to trisect an angle of $60°$. Then $\cos 60° = \frac{1}{2}$. If $x = \cos\theta$, we have $4x^3 - 3x = \frac{1}{2}$ or $8x^3 - 6x + 1 = 0$. Let $y = 2x$. Then $y^3 - 2y + 1 = 0$. Show this equation is irreducible. Therefore angles cannot in general be trisected and a regular 9-gon cannot be constructed.

Level 2

1. Construction of a regular n-gon can be simplified by introduction of complex numbers. Let the centre be $(0, 0)$ and the radius 1. Then if a point x, y can be constructed the number $x + iy$ can be obtained by repeated extraction of square roots. The converse is also true. Let one vertex be $(1, 0)$. The next is $(\cos\frac{2\pi}{n}, \sin\frac{2\pi}{n})$. Let $z = \cos\frac{2\pi}{n} + i\sin\frac{2\pi}{n}$. Then it can be shown $z^n = 1$. Factor $z^n - 1$ to prove z satisfies the equation $z^{n-1} + z^{n-2} + \ldots + z + 1 = 0$.
2. Prove if regular m, n-gons can be constructed, and m and n are relatively prime so can a regular mn-gon. We can construct angles of $\frac{2\pi}{m}, \frac{2\pi}{n}$. Show there exist integers r, s such that $\frac{2\pi r}{m} - \frac{2\pi s}{n} = \frac{2\pi}{mn}$.
3. Prove by repeated bisection of an angle we can construct a regular 2^n-gon.
4. Prove a regular 5-gon can be constructed, by solving $z^4 + z^3 + z^2 + z^1 + z^0 = 0$. Write $u = z + z^{-1}$. Express $0 = z^2 + z + 1 + z^{-1} + z^{-2}$ as a sum $u^2 + u + c = 0$. Thus u satisfies a quadratic. Show that z satisfies a quadratic involving u.

5. For n prime show $z^{n-1} + z^{n-2} + \ldots + z^1 + z^0$ is irreducible using the Eisenstein Irreducibility Criterion. Let $z = u + 1$. Then this polynomial is

$$\frac{(u+1)^n - 1}{u}$$

Show all coefficients after u^n are divisible by n but the constant term is precisely n. Therefore a regular p-gon is not constructible by ruler and compass for p not of the form $2^k + 1$.

6. Prove the Eisenstein Irreducibility Criterion. If $f(x) = g(x)h(x)$, prove that all coefficients of g, h except the first are divisible by p by reducing f, g, h modulo p. Then show the last nonzero coefficient of f is divisible by p^2, contradiction.

Level 3

1. Derive the solution of the general cubic equation, as follows. We use solvability of S_3 by the series $\{e\} \subset \mathbf{Z}_3 \subset S_3$. Let the roots be x_1, x_2, x_3. Begin with $\Delta = (x_1 - x_2)(x_2 - x_3)(x_1 - x_3)$. Since Δ^2 is a symmetric function we can express it in terms of the coefficients c_1, c_2, c_3 of $x^3 - c_1 x^2 + c_2 x - c_3$ by the methods of Theorem 5.5.5. Assume that $c_1 = 0$. This yields Δ^2.
2. Let ω be a nontrivial root of $\omega^2 = 1$ where

$$\omega = \frac{1 - \mathrm{i}\sqrt{3}}{2}.$$

Show $y = x_1 + \omega x_2 + \omega^2 x_3$ is sent to a multiple of itself by a power of ω under the cyclic permutation $x_1 \to x_2 \to x_3$ and $x_1 \to x_2 \to x_3$. Hence y^3 is invariant under $\mathbf{Z}_3$. So it belongs to the same group as Δ. Compute y^3 and write it as a function of c_1, c_2, c_3, Δ.

3. By symmetry write an expression for $z = x_1 + \omega^2 x_2 + \omega x_3$.
4. Since $x_1 + x_2 + x_3$, $x_1 + \omega x_2 + \omega^2 x_3$, $x_1 + \omega^2 x_2 + \omega x_3$ are known and

$$\begin{bmatrix} 1 & 1 & 1 \\ 1 & \omega & \omega^2 \\ 1 & \omega^2 & \omega \end{bmatrix}$$

is nonsingular, explain how x_1, x_2, x_3 may be computed by solving simultaneous linear equations.

5. If $z^{p^2} = 1$ but $z^p \neq 1$, show z satisfies $g(z) = (z^p)^{p-1} + (z^p)^p + \ldots + z^p + z^0 = 0$. Show that this equation is irreducible. Suppose not. Use the fact that if z is a root, so is $z\omega$ for any root ω of $\omega^p = 1$, and irreducibility of $z^{p-1} + z^{p-2} + \ldots + z^1 + z^0$. Given the minimum polynomial $f(z)$ of z, add $f(z) + f(\omega z) + \ldots + f(\omega^{p-1} z)$ to get a polynomial involving only

powers of z^p. Then $f(x)|g(x)$ since $g(z)=0$. Yet this is not possible since $y^{p-1}+y^{p-2}+\ldots+y+1$ is irreducible. So for $p>2$, a regular polygon of degree p^2 cannot be constructed with ruler and compass.

6.3 FINITE FIELDS

In this section we will study fields having a finite number of elements. All such fields are known and they are used in many combinatorial constructions, for instance in coding theory and the design of Latin squares. The fields $\mathbf{Z_p}$ are examples of finite fields, and all other finite fields turn out to be extension fields of some field $\mathbf{Z_p}$.

THEOREM 6.3.1. *Every finite field* $\mathbf{F}$ *has* p^n *elements for some prime* p, *and* $\mathbf{Z_p}$ *is a subfield of* $\mathbf{F}$.

Proof. Let $\mathbf{F}$ be a finite field. Then the additive group is finite, so for some integer n, a sum of n copies of 1 is zero. We write $n1=0$. Choose n to be the least such positive integer. Suppose $n=ab$ where $a, b \in \mathbf{Z}^+, 1<a, b<n$. Then $(a1)(b1)=n1$ but $a1\neq 0$ and $b1\neq 0$. This contradicts the properties of a field. So n is a prime p, unless $n=1$. But if $n=1$ then $1=0$, which is false. So assume $p>1$.

The set S of all sums of copies of 1 will be closed under addition and multiplication. The additive and multiplicative groups of $\mathbf{F}$ are finite. This means that some repeated sum of an element will be its additive inverse. The same is true for products. So any subset closed under addition and multiplication also contains all additive inverses and multiplicative inverses of nonzero elements. So it is a subfield.

The mapping $a \to a1$ is a ring homomorphism from $\mathbf{Z}$ onto S. Its kernel is precisely the set of multiples of p. So by Theorem 4.3.5, the quotient ring $\mathbf{Z_p}$ is isomorphic to S.

Now $\mathbf{F}$ is a vector space over S. So $\mathbf{F}$ has a basis $x_1, x_2, \ldots, x_k$. Then $\mathbf{F}$ is in 1-1 correspondence with the set of all sums $a_1x_1+a_2x_2+\ldots+a_kx_k$ where $a_1, a_2, \ldots, a_k \in \{\overline{0}, \overline{1}, \overline{2}, \ldots, \overline{p-1}\}$. The number of such sums is p^k. So $|\mathbf{F}|=p^k$. □

We note that the integer $p = \text{char}(\mathbf{F})$.

THEOREM 6.3.2. *In any field no polynomial equation of degree* n *can have more than* n *distinct roots.*

Proof. If r is a root of $f(x)$ then $f(x)=(x-r)g(x)+a$ where a is 0 or has degree 1. Setting $x=r$ we find $a=0$. So $f(x)=(x-r)g(x)$. Thus we can factor off an $x-r$ for each r. A polynomial of degree n cannot have more than n factors of the form $(x-r)$. □

We showed in Theorem 4.4.5 that any finite subgroup of the multiplicative group of a field is cyclic. Therefore if $\mathbf{F}$ is a field of order p^n its nonzero elements are isomorphic to $\mathbf{Z}_{pn-1}$.

Finite fields are usually constructed as quotient rings

$$\frac{\mathbf{F}(x)}{p(x)\mathbf{F}(x)}$$

where $p(x)$ is an irreducible polynomial over $\mathbf{Z}_\mathbf{p}$. The elements $1, x, x^2, \ldots, x^{n-1}$ form a basis if $p(x)$ has degree n. This determines the addition, for example in an extension of $\mathbf{Z}_5$ of degree 2:

$$(2+3x)+(4+x) \equiv (2+4)+(3+1)x \equiv 6+4x \equiv 1+4x \pmod 5$$

Multiplication is determined by multiplying two polynomials and then using $p(x)$ as a relation to reduce the powers of x. For instance,

$$(2+3x)(4+x) = 6+12x+2x+3x^2 = 1+4x+3x^2 =$$

$$1+4x+3(2) = 7+4x = 2+4x$$

provided that $p(x) = x^2 - 2$ so that $x^2 = 2$.

EXAMPLE 6.3.1. Over $\mathbf{Z}_2$ the polynomial $x^2 + x + 1$ has no roots, hence no factors of degree 1 and is therefore irreducible.

In the quotient ring there are four elements $0, 1, \bar{x}, 1+\bar{x}$. Addition and multiplication are described below.

+	0	1	$\bar{x}$	$1+\bar{x}$
0	0	1	$\bar{x}$	$1+\bar{x}$
1	1	0	$1+\bar{x}$	$\bar{x}$
$\bar{x}$	$\bar{x}$	$1+\bar{x}$	0	1
$1+\bar{x}$	$1+\bar{x}$	$\bar{x}$	1	0

$\times$	0	1	$\bar{x}$	$1+\bar{x}$
0	0	0	0	0
1	0	1	$\bar{x}$	$1+\bar{x}$
$\bar{x}$	0	$\bar{x}$	$1+\bar{x}$	1
$1+\bar{x}$	0	$1+\bar{x}$	1	$\bar{x}$

For instance $\bar{x}^2 \equiv -1 - \bar{x} \equiv 1 + \bar{x} \pmod 2$. And $\bar{x}(1+\bar{x}) = \bar{x} + \bar{x}^2 = \bar{x} + 1 + \bar{x} = 1$.

Another way to deal with the multiplicative structure of a finite field is to construct a logarithm table. That is, find an element g which is a generator of the cyclic group of the field, and write out all powers of g. Then two polynomials can rapidly be multiplied simply by adding the exponents of g. We illustrate this process for a field of order 8.

EXAMPLE 6.3.2. We find a logarithm table for the field of order 8 generated by z such that $z^3 + z + 1 = 0$. The powers of z are:

$z^0 = 1$
$z^1 = z$
$z^2 = z^2$
$z^3 = 1 + z$
$z^4 = z(1+z) = z + z^2$
$z^5 = z(z + z^2) = z^2 + z^3 = z^2 + 1 + z$
$z^6 = z(z^2 + 1 + z) = z^3 + z + z^2 = 1 + z + z + z^2 = 1 + z^2$
$z^7 = 1$

For instance $(z + z^2)(z^2 + 1 + z) = z^4 z^5 = z^9 = z^2$ since $z^7 = 1$.

The major theorem about finite fields is that there exists one and only one field of order p^n for every prime p and positive integer. n. To show existence it suffices to show there exists at least one irreducible polynomial of degree n, then the quotient ring has degree n over $\mathbf{Z_p}$.

LEMMA 6.3.3. *Let $\mathcal{R}$ be a commutative ring with 1, in which $p1 = 0$ for a prime p. Then for any $r > 0$ the mapping $x \to x^{p^r}$ is a ring homomorphism, which is the identity on $n1$ for any n.*

Proof.

$$(x + y)^p = x^p + y^p + \sum_{r=1}^{p-1} \binom{p}{r} x^r y^{p-r}.$$

But for $r = 1$ to $p - 1$, $p | p!$ but not $r!(p - r)!$, since p does not divide any of the factors of the latter expression. Thus $p | \binom{p}{r}$. So in $\mathcal{R}$, $(x + y)^p = x^p + y^p$. Also, by commutativity $(xy)^p = x^p y^p$. And by Corollary 4.4.4, $(n1)^p = n1$. Therefore $x \to x^p$ is a ring homomorphism which is the identity on $\{n1\}$. But $x \to x^{p^r}$ is the r-fold composition of $x \to x^p$ with itself. So it is also a ring homomorphism which is the identity on $\{n1\}$. □

EXAMPLE 6.3.3. The mapping $x \to x^p$ is an automorphism of any field of order p^n.

In the ring of polynomials over any field we can define a derivative by $D(c_0 x^n + c_1 x^{n-1} + \ldots + c_n) = nc_0 x^{n-1} + (n-1)c_1 x^{n-2} + \ldots + c_{n-1}$. Then the derivative is linear: $D(af + bg) = aDf + bDg$. The formula $D(fg) = D(fg) = fDg + gDf$ holds for any powers x^n, x^m of x. By linearity one can then show it holds for any f, g.

LEMMA 6.3.4. *The polynomial $x^{p^n} - x$ over a field in which $p1 = 0$ has no factors which are squares of a polynomial of positive degree.*

Proof. $D(fg^2) = fDg^2 + g^2Df = fgDg + fgDg + g^2Df$. So if g^2 divides a polynomial then g divides its derivative. But the derivative of $x^{p^n} - x$ is $0 - 1 = -1$. □

THEOREM 6.3.5. *The polynomial $x^{p^n-1} - 1$ modulo p has no irreducible factors of degree larger than n, but has some irreducible factor of degree n.*

Proof. Suppose this equation has an irreducible factor $f(x)$ of degree larger than n. Then form the quotient ring R_Q associated with $f(x)$. This quotient ring has as a basis $1, x, x^2, \ldots, x^{k-1}$ and is a field of order p^k where $k > n$. Since $f(x)\ x^{p^n-1} - 1$, in this field $x^{p^n-1} = 1$. So $x^{p^n} = x$. But this implies for any k, $(x^k)^{p^n} = x^k$. By Lemma 6.3.4,

$$(a_1x^{k-1} + a_2x^{k-2} + \ldots + a_n1)^{p^n} = a_1x^{k-1} + a_2x^{k-2} + \ldots + a_11.$$

So every element of R_Q satisfies the equation $y^{p^n} = y$. This contradicts that fact that in a field a polynomial of degree p^n cannot have $p^k > p^n$ roots. So no irreducible factors of degree greater than n can occur.

Suppose that $x^{p^n} - x$ divides some power of $x(x^p - x)(x^{p^2} - x) \ldots (x^{p^{n-1}} - x)$. Then by Lemma 6.3.4 and unique factorization of polynomials,

$$x^{p^n} - x \mid x(x^p - x)(x^{p^2} - x) \ldots (x^{p^{n-1}} - x)$$

But since

$$p^n > 1 + p + p^2 + \ldots + p^{n-1} = \frac{p^n - 1}{p - 1}$$

that is impossible. So let $f(x)$ be an irreducible factor of $x^{p^n} - x$ which does not divide $x(x^p - x)(x^{p^2} - x) \ldots (x^{p^{n-1}} - x)$. The degree of $f(x)$ cannot be greater than n by Theorem 6.3.2. It will suffice to show it has degree exactly n.

Suppose $f(x)$ has degree $k < n$. The quotient

$$\frac{\mathbf{Z_p}[x]}{f(x)\mathbf{Z_p}[x]}$$

has order p^k and in it x satisfies $x^{p^k} = x$ since the multiplicative group has order $p^k - 1$. Yet also x satisfies $f(x) = 0$, which is its minimum polynomial in the quotient field. Therefore $f(x) \mid x^{p^k} - x$. But this is contrary to the assumption on $f(x)$. □

EXAMPLE 6.3.4. For $p = 3$, $n = 2$ we factor $x^{3^2-1} - 1 = x^8 - 1$. It is divisible by $x^{3-1} - 1 = x^2 - 1$. The quotient is $x^6 + x^4 + x^2 + 1$. This factors as $(x^2 + 1)(x^4 + 1)$. The factor $(x^4 + 1)$ equals $(x^2 - x - 1)(x^2 + x - 1)$. Any of the factors $x^2 + 1$, $x^2 - x - 1$, $x^2 + x - 1$ is an irreducible polynomial giving a field of order 9. They in fact give the same field.

THEOREM 6.3.6. *For any prime p and positive integer n, there exists a unique field of order p^n. A field of order p^m is isomorphic to a subfield of a field of order p^n if and only if $m \mid n$.*

Proof. Existence of a field of order p^n follows from Theorem 6.3.5 together with Theorem 6.1.2.

Let $\mathbf{F}_1$ and $\mathbf{F}_2$ be two fields of order p^n. Then in each field all elements satisfy $x^{p^n} - x$, and this polynomial factors into linear factors.

Let $f(x)$ be an irreducible factor (over $\mathbf{Z}_p$) of this polynomial having degree n. Then some linear factors $(x - r)$, $r \in \mathbf{F}_1$ and $(x - s)$, $s \in \mathbf{F}_2$, divide $f(x)$ by unique factorization. So $f(r) = 0$ in $\mathbf{F}_1$ and $f(s) = 0$ in $\mathbf{F}_2$.

We next observe that the powers $1, r, r^2, \ldots, r^{n-1}$ are linearly independent over $\mathbf{Z}_p$. Suppose not. Then

$$a_1 r^{n-1} + a_1 r^{n-2} + \ldots + a_n = 0$$

where not all a_i are 0. Call this polynomial $g(x)$. Then the degree of $g(x)$ is less than n, so $g(x)$ is not divisible by $f(x)$. Since $f(x)$ is irreducible, g.c.d. of $f(x), g(x)$ is 1. Therefore $f(x)u(x) + v(x)g(x) = 1$. Put $x = r$. Then $1 = 0$. This proves $1, r, r^2, \ldots, r^{n-1}$ are linearly independent. Their span has p^n elements. So it is all of $\mathbf{F}_1$. So they are a basis. Likewise $1, s, s^2, \ldots, s^{n-1}$ are a basis for $\mathbf{F}_2$.

Now define a function $h: \mathbf{F}_1 \to \mathbf{F}_2$ by

$$h(a_n + a_{n-1}r + \ldots + a_1 r^{n-1}) = a_n + a_{n-1}s + \ldots + a_1 s^{n-1}.$$

Then h is an isomorphism of vector spaces, since $1, r, r^2, \ldots, r^{n-1}$ and $1, s, s^2, \ldots, s^{n-1}$ are bases.

Since $f(r) = 0$, each power $r^n, r^{n+1}, \ldots$ can be expressed as a linear combination of $1, r, r^2, \ldots, r^{n-1}$ using the coefficients of f. And $s^n, s^{n+1}, \ldots$ will be the same linear combinations of $1, s, s^2, \ldots, s^{n-1}$ since $f(s) = 0$. Therefore $h(r^j) = s^j$ for all $j > 0$. Therefore for any polynomial $p(r)$, $h(p(r)) = p(s)$. Thus $h(p_1(r)p_2(r)) = p_1(s)p_2(s) = h(p_1(r))h(p_2(r))$. Therefore h is an isomorphism of rings from $\mathbf{F}_1$ to $\mathbf{F}_2$.

Suppose $\mathbf{F}_1$ of order p^m is a subfield of $\mathbf{E}$. Then $\mathbf{E}$ is a vector space over $\mathbf{F}_1$. Let a basis be $x_1, x_2, \ldots, x_k$. Therefore every element of $\mathbf{E}$ is a unique linear combination $a_1x_1 + a_2x_2 + \ldots + a_k x_k$, $a_i \in \mathbf{F}_1$. So $\mathbf{E}$ has $(p^m)^k$ elements.

Conversely suppose $mk = n$ for some k. Then $x^{p^m} - x \mid x^{p^n} - x$. Let $\mathbf{E}$ be a field of order p^n. Let $f(x)$ be an irreducible factor of $x^{p^m} - x$ of degree

m. Then for some $r \in \mathbf{E}, x - r \mid f(x)$. So $f(r) = 0$. As above, we can show $1, r, r^2, \ldots, r^{m-1}$ generate a subfield of $\mathbf{E}$ having p^m elements. □

Finite fields are also called ***Galois fields***. A finite field of order p^n is denoted $GF(p^n)$.

Many methods we have used here can also be used in the study of ***infinite fields***.

EXERCISES

Level 1

1. Write out the addition table of a field of order 9. The 9 elements have the form $a + bx$, $a, b \in \overline{0}, \overline{1}, \overline{2}$ and are added as polynomials in $\mathbf{Z}_3$.
2. Work out the multiplication table of this field. Use $x^2 + 1$ as a generating polynomial, so that $x^2 = -1$ in this field. (We can write the elements as $a + b\mathrm{i}, \mathrm{i} = \sqrt{-1}$ instead of $a + bx$.)
3. Find a multiplication generator g of this field and compute its powers $g, g^2, \ldots, g^8$.
4. Use the table in Exercise 3 to find the fourth power of $(1 + x)$.
5. Find the inverse of $(1 - x)$ in the field of order 9.
6. In the field of order 8 use the logarithm table given in Example 5.3.2 to multiply $(1 + z)$ and $(1 + z^2)$.
7. In the field order 4 describe the automorphism $x \to x^2$. Thus $x, 1 + x$ are symmetrical.
8. Find an irreducible polynomial of order 2 over the field of order 9 having the form $x^2 - a$, that is, a is not a square of an element.

Level 2

1. Find an irreducible monic polynomial of degree 3 over $\mathbf{Z}_3$. It must satisfy $f(0) \neq 0, f(1) \neq 0, f(-1) \neq 0$ modulo 3.
2. Show that if an irreducible polynomial of degree n, $p > n > 1$ exists over $\mathbf{Z}_\mathbf{p}$ one exists having the coefficient of x^{n-1} equal to zero.
3. In a field of order 27 construct the multiplication table of $1, x, x^2$.
4. Show that for any odd prime p there exists an irreducible quadratic over $\mathbf{Z}_\mathbf{p}$ of the form $x^2 - a$. Use results on quadratic residues from Proposition 4.4.7 and Corollary 4.4.8.
5. Show that for any odd prime p of the form $(6n + 1)$ there exists an irreducible cubic of the form $x^3 - a$ over $\mathbf{Z}_\mathbf{p}$.
6. Find an irreducible polynomial of degree 2 over $GF(4)$. Using it find a logarithm table for the field of order 16.

Level 3

1. Show the automorphism $x \to x^p$ generates a cyclic group $\mathbf{Z}_\mathbf{n}$ of automorphisms of a field of order p^n.

2. Show every automorphism of a finite field has the form $x \to x^{p^k}$.
3. If $f(x)$ is a monic polynomial with coefficients in $\mathbf{Z}$ which factors over $\mathbf{Z}$ it factors over $\mathbf{Z_p}$ for every prime p. Do you think the converse is true? (This is a deep question. For quadratics it holds by quadratic reciprocity.)
4. For 2, 3 it has been remarked that $x^k + x + 1$ is irreducible over $\mathbf{Z}_2$. What about 4, 5? Try irreducible quadratic factors $x^2 + ax + 1$.
5. Show that every irreducible polynomial of degree n is a divisor of $x^{p^n} - x$ which does not divide $x^{p^k} - x$ for $k < n$. Hence factoring $x^{p^n} - x$ is not in general a good method of finding irreducible polynomials.

6.4 CODING THEORY

Coding theory is concerned with methods of symbolizing data such that most errors can be detected. This is done because correctly coded messages have a certain form. For instance if the first digit of a number could not exceed 7, and a 9 is received there must have been an error. Error correcting codes cannot correct all possible errors since for any correct message any other correct message of the same length could have been sent instead. They correct errors occurring in only a small proportion of the digits in each group of numbers. Such codes are widely used in computer circuitry and in digital radio communications.

The simplest error-detecting code of any interest is a parity check. Suppose a coded message is in blocks of 7 digits 0, 1. Then an extra digit is added to each block. This digit is chosen to make the total number of ones in the block even.

EXAMPLE 6.4.1. To the block 1 1 1 0 1 0 1 is added a digit 1, giving 1 1 1 0 1 0 1 1 having an even number of 1 digits.

This code detects any single error in a block. If a single error is made, the number of 1 digits is changed by 1 and therefore becomes odd. If a receiver sees a block having an odd number of 1 digits, he knows an error has occurred. However, if exactly 2 errors occur, the code cannot detect any error.

This is an error-detecting code. There also exist error-correcting codes which can tell what the correct block should have been, provided that 1 or more errors occurred. A trivial example is the code with blocks of length 3 each all zeros or all ones, 0 0 0 or 1 1 1. If a single error is made, the receiver can tell whether the original block was 1 1 1 or 0 0 0 by looking at whether there is a majority of '0's or a majority of '1's.

To be precise we assume the message is encoded in blocks of m digits from a set S containing q possible digits. Each block is encoded as a block of n digits from S where $n > m$.

EXAMPLE 6.4.2. For the parity check $m = 7$, $q = 2$, $n = 8$. The set S of possible digits is $\{0, 1\}$. A block of length 7 from $\{0, 1\}$ is encoded as a block of length 8 by adding a parity check.

A code involves a coding function from length m sequences from the set S of symbols to length n sequences, and then a decoding function to recover the original. However, for error detecting and correction the essential thing is the set of encoded words, not the coding function: any 1–1 function may be chosen. The coding and decoding functions should, however, be rapidly computable.

We assume q is a power of a prime and regard S as $GF(q)$. Also we regard a sequence of length m as a vector of length m over $GF(q)$.

DEFINITION 6.4.1. Let $V(n, q)$ denote the set of n-dimensional vectors over $GF(q)$. A *code* is a subset of $V(n, q)$. A *coding function* is a 1–1 function $c: V(m, q) \subset V(n, q)$ whose image lies in the code. A *decoding function* is a function d such that $d(c(v)) = v$ for all $v \in V(n, q)$.

EXAMPLE 6.4.3. For the case of a parity check, we have a function $f_E: V(m, 2) \to V(m + 1, 2)$ such that

$$f_E(v_1, v_2, \dots, v_n) = (v_1, v_2, \dots, v_n, v_1 + v_2 + \dots + v_n)$$

DEFINITION 6.4.2. For two vectors $u = (u_1, u_2, \dots, u_n)$, $v = (v_1, v_2, \dots, v_n)$ the *Hamming distance* $H(u, v)$ is the number of places where the vectors differ, i.e. $|\{i : u_i \neq v_i\}|$.

EXAMPLE 6.4.4. The Hamming distance $H((0\,1\,1\,0\,1), (1\,1\,1\,0\,0))$ is 2, since these vectors differ in two locations.

Note that $H(u, v) = H(u - v, 0)$. The function $H(u, 0)$ is sometimes called the *weight* of u, for any vector u.

THEOREM 6.4.1. *A code can detect up to k errors if and only if the Hamming distance between two encoded words is at least k + 1. It can correct up to k errors if and only if the Hamming distance between any two encoded words is at least 2k + 1.*

Proof. We will prove the latter statement first. Suppose the distance between any two encoded words is at least $2k + 1$. Suppose that an encoded word v is sent and that after at most k errors in transmission it becomes w. Then we claim that v is the unique encoded word such that $H(v, w) \leqslant k$. Suppose for $u \neq v$, $H(u, w) \leqslant k$. Then $H(u, v) \leqslant 2k$. But this is contrary to assumption. So the receiver can find the unique encoded word w such that $H(v, w) \leqslant k$. This must be the correct original word v.

Conversely suppose the Hamming distance between two coded words a, b is $H(a, b) \leqslant 2k$. Form a word z intermediate between the two, which agrees with both if both agree, else agrees with a in k places where they differ and with b in the other $H(a, b) - k$. Then $H(z, a) = H(a, b) - k \leqslant k$ and $H(z, b) = k$ so z could have arisen from either a or b.

If k errors occur in a word a we obtain b with $H(a, b) = k$. So b is not a code word under the assumption of the first statement, and the error can be detected. If two words have Hamming distance $\leqslant k$ then one could be erroneously received as the other. So the code cannot always detect k errors. □

The problem is to find codes which are efficient. If n is twice m, then the code takes up 100% more time than the original transmission. So n should be not too much larger than m.

DEFINITION 6.4.3. The *rate* of a code is $\frac{n}{m}$.

EXAMPLE 6.4.5. The example 000, 111 used to encode 0, 1 has rate $\frac{3}{1} = 3$. It is not very efficient.

One major problem in coding theory is minimizing the rate (another is finding coding and decoding algorithms which run quickly on a computer). The rate is studied by various inequalities, many quite difficult to prove.

The following inequality is called the ***Hamming bound***. It gives an upper bound on n for given n, q, t where t is the number of errors.

PROPOSITION 6.4.2. *If a code can correct up to t errors, then*

$$q^{n-m} \geqslant \binom{n}{0} + (q-1)\binom{n}{1} + (q-1)^2\binom{n}{2} + \ldots + (q-1)^t\binom{n}{t}$$

Proof. The number of vectors at Hamming distance k from a given vector v is $\binom{n}{k}(q-1)^k$ since we can choose k locations out of n for the errors and then choose the new values in each case by $(q-1)$ possibilities. Thus the right-hand side is the number of vectors of Hamming distance less than or equal to t from a given vector.

For a code which can correct up to t errors, consider the set of vectors within Hamming distance t of an encoded vector. The encoded vector can be chosen in q^m ways and then the vectors with errors in

$$\binom{n}{0} + (q-1)\binom{n}{1} + \ldots + (q-1)^t\binom{n}{t}$$

ways. All these choices will be distinct by the proof of Theorem 6.4.1.

So we have this many distinct vectors:

$$q^m\left(\binom{n}{0} + (q-1)\binom{n}{1} + \ldots + (q-1)^t\binom{n}{t}\right)$$

But there are only q^n vectors in all in $V(n, q)$ so this quantity does not exceed q^n. This implies the inequality. □

Codes in which this inequality is an equality are said to be ***perfect codes,*** and are of special interest.

There exist a family of perfect 1-error correcting codes, where n is of the form

$$\frac{q^k - 1}{q - 1}$$

called ***Hamming codes.*** We consider these for $q = 2$.

DEFINITION 6.4.4. Let A be the $(2^k - 1) \times k$ matrix such that the jth row of A is the number j written in binary notation.

EXAMPLE 6.4.6. For $k = 3$, A is the matrix

$$\begin{bmatrix} 0 & 0 & 1 \\ 0 & 1 & 0 \\ 0 & 1 & 1 \\ 1 & 0 & 0 \\ 1 & 0 & 1 \\ 1 & 1 & 0 \\ 1 & 1 & 1 \end{bmatrix}$$

To code a (0, 1)-vector v of length $2^k - k - 1$, form a vector w of length $2^k - 1$ such that v is the sequence of entries in w in locations other than the $1, 2, 4, \ldots, 2^{k-1}$ components of w.

These k components are now chosen so that $wA = 0$.

EXAMPLE 6.4.7. Suppose we wish to encode the vector $v = (1, 1, 0, 1)$. Write $w = (w_1, w_2, 1, w_4, 1, 0, 1)$, where the components of w are the unknown w_{2^j} together with the entries of v.

$$wA = [w_4 \; w_2 \; w_1 + 1]$$

So let $w_4 = 0$, $w_2 = 0$, $w_1 = 1$. The encoded vector w is $(1, 0, 1, 0, 1, 0, 1)$.

PROPOSITION 6.4.3. *For any vector u there exists a unique vector w such that* $wA = 0$ *and* $w_i = u_i$ *for i not of the form* 2^j.

Proof. The equation that $wA_j = 0$ is

$$\sum_{h \in S} w_h = 0$$

here S is the set of numbers 1 to $2^k - 1$ whose jth digit in binary notation is 1. Thus this equation has the form

$$w_{2^k-j} + \sum_{d \in S} w_d = 0$$

where the numbers d are not powers of 2. Therefore $w_d = u_d$.

The unique solution is

$$w_{2^k-j} = \sum_{d \in S} u_d$$

$w_i = u_i$ for i not of the form 2^j. □

This defines the ***encoding function*** f_E.

THEOREM 6.4.4. *The Hamming code can correct any 1-digit error.*

Proof. Suppose that the correctly coded vector is v, and that an error in 1-digit has been made, resulting in a vector w. Then $v - w$ is a vector u having exactly one 1. The location of this 1 is the location of the error. Therefore

$$wf_E = (v + u)f_E = 0 + uf_E = uf_E$$

But if u has a 1 in place j, then uf_E is the jth row of f_E. But this row is the number j written in binary notation. So by reading wf_E in binary notation we have j, the location of the error. By changing this digit we have corrected the error. □

To decode a message in Hamming code, simply delete all digits w_{2^j}.

It has been proved that the only perfect codes are the Hamming codes, the trivial codes where $m = 1$, $n = 2t + 1$ which repeat a digit $2t + 1$ times, a code with $q = 2$, $n = 23$, $m = 12$, $t = 3$, and a code with $q = 3$, $n = 11$, $m = 6$, $t = 2$, both due to Golay. The latter are associated with special permutation groups called ***Mathieu groups.***

C. Shannon proved that essentially random codes for large enough blocks sizes, at a fixed probability ρ of error per digit achieve arbitrarily low probability ϵ of error per digit and rates arbitrarily close to the maximum

$$1 + \rho \log_2 \rho + (1 - \rho) \log_2 (1 - \rho)$$

Another bound in coding theory is that of Joshi.

THEOREM 6.4.5. *A code which can detect t errors has $m \leqslant n - t$.*

Proof. Let C be a code any pair of whose members have Hamming distance at least $t + 1$. Let V be the set of vectors having zeros in places $t + 1, t + 2, \ldots, n$. Then all words $c + v$, $c \in C$, $v \in V$ are distinct, else if $c_1 + v_1 \subset c_2 + v_2$ then

$c_1 - c_2 = v_1 - v_2 \in V$ so c_1 and c_2 have Hamming distance $\leqslant t$. This is a contradiction. So these $q^m \cdot q^t$ words are distinct in $V(n, q)$. So $q^{m+t} \leqslant q^n$. □

This proof used the fact that V is a set of words of weight at most t, closed under addition.

EXERCISES

As in the preceding section, here m is the length of a message block before encoding, q the number of possible symbols, n the length of a message block after encoding and t the number of errors correctible.

Level 1

1. Define a modulo 3 code by adding a digit such that the sum of all digits is divisible by 3. Encode 0 1 2 1 1.
2. Show the preceding code detects a single error.
3. Encode 1 0 0 0 using the Hamming code.
4. Decode 1 0 1 1 0 1 1 in the Hamming code.
5. If a message 1 1 1 1 1 1 1 in the Hamming code is received, how many errors occurred (assuming at most 1 did)? Where?
6. What are the rates of the Hamming codes?
7. Prove the Gilbert–Varshamov lower bound. There exists a t error-correcting code over $GF(q)$ with μ code words and

$$\sum_{i=0}^{2t} \binom{n}{i} (q-1)^i \quad \mu \geqslant q^n$$

Assume that μ code words have been found having Hamming distance at least $2t + 1$ from the rest. Show that if the inequality is false then words at Hamming distance less than or equal to $2t$ from existing words do not exhaust all words. So a new one can be added, increasing μ.

Level 2

1. Construct a nonperfect code with $n = (2t + 1)m$, $m > 1$ by repeating each block t times. Prove it corrects t errors.
2. Characterize the set of correctly coded words in the Hamming code.
3. Define the Hamming codes for $q > 2$.
4. For $q > 2$ prove results analogous to those given in the text for $q = 2$.
5. Give examples of encoding in the Hamming codes for $q > 2$.
6. Give examples of decoding in the Hamming codes for $q > 2$.
7. For a perfect code to exist

$$\sum_{i=0}^{t} \binom{n}{i} (q-1)^i$$

must be a power of q. For $t = 2$ give examples of n, q which are ruled out.

8. Fix t, q. Suppose a perfect code existed for n. Show that as $n \to \infty$ the rate would approach 1.

Level 3

1. Prove a code is perfect if and only if every sequence of length m can be decoded by the procedure of Theorem 6.4.1.
2. What is the average Hamming distance between two words of length n? This is the same as the average number of nonzero entries of a given word.
3. Consider a code satisfying the Gilbert-Varshamov lower bound. If $\frac{t}{n} \to 0$ does the rate approach 1?
4. Look up and write out in your own words a proof of Shannon's theorem.

6.5 CYCLIC CODES

A code is ***linear*** if it is a subspace of $V(n, q)$. It is ***cyclic*** if in addition any cyclic rearrangement of a code word is one. Cyclic codes turn out to be ideals in $V(n, q)$ if it is regarded as the ring

$$\frac{\mathbf{F}[x]}{(x^n - 1)\mathbf{F}[x]}$$

Such ideals can be dealt with by specifying a generating polynomial, which can be chosen to divide $(x^n - 1)$.

Three classes of such codes are the ***BCH codes***, constructed by R.C. Bose and D.K. Ray-Chaudhuri, and independently by A. Hoquenghem, in 1959, the ***Reed-Solomon codes*** of I.S. Reed and G. Solomon, and the ***Fire codes*** of P. Fire.

PROPOSITION 6.5.1. *A code is cyclic if and only if it is an ideal in*

$$\frac{\mathbf{F}[x]}{(x^n - 1)\mathbf{F}[x]}$$

Proof. A linear subspace in this ring is an ideal if and only if it is preserved under multiplication by x. But this takes $a_0 + a_1 x + \ldots + a_{n-1} x^{n-1}$ to $a_{n-1} + a_0 x + a_1 x + \ldots + a_{n-2} x^{n-1}$ which is a cyclic rotation. □

PROPOSITION 6.5.2. *Any ideal in*

$$\frac{\mathbf{F}[x]}{(x^n - 1)\mathbf{F}[x]}$$

contains a unique monic polynomial of lowest degree. This polynomial generates the ideal. It divides $(x^n - 1)$.

Proof. An ideal $\mathcal{I}$ in

$$\frac{\mathbf{F}[x]}{(x^n - 1)\mathbf{F}[x]}$$

gives an ideal I, in $\mathbf{F}(x)$, namely all polynomials whose image lies in I. If I_1 had two monic polynomials of lowest degree, their difference would give a monic polynomial $g(x)$ of lower degree, which is a contradiction. The g.c.d. of this polynomial and $(x^n - 1)$ lies in I_1 so this g.c.d. cannot have degree less than $g(x)$. So it must be $g(x)$. So $g(x)|x^n - 1$. If $g(x)$ did not divide any member of I there would be a similar contradiction. □

The problem is to find an ideal which has a large number of elements and can correct or detect many errors. The number of errors detectible is the minimum Hamming distance which equals the minimum weight of any polynomial. The number of members of the ideal is $q^{n-\deg(g)}$ since the quotient ring has basis $1, x, \ldots, x^{\deg(g)-1}$, and so has $q^{\deg(g)}$ elements.

The following lemma is the means of guaranteeing that polynomials of small weight are not in the ideal.

LEMMA 6.5.3. *The determinant of*

$$\begin{bmatrix} z_1 & z_1^2 & \cdots & z_1^n \\ z_2 & z_2^2 & \cdots & z_2^n \\ & \cdots\cdots & & \\ z_n & z_n^2 & \cdots & z_n^n \end{bmatrix}$$

equals $(z_1 z_2 \ldots z_n) \prod_{i>j} (z_i - z_j)$.

Proof. For $n = 2$ this result is correct. Assume it is true for all numbers less than n. Subtract z_n times column $n-1$ from column n, z_n times column $n-2$ from column $n-1$, and so on. This will not affect the determinant. We have

$$\begin{bmatrix} z_1 & z_1(z_1 - z_n) & \cdots & z_1^{n-1}(z_1 - z_n) \\ z_2 & z_2(z_2 - z_n) & \cdots & z_2^{n-1}(z_2 - z_n) \\ & & \cdots\cdots & \\ z_n & 0 & \cdots & 0 \end{bmatrix}$$

Expand by minors on the last row, and factor out the quantities in parentheses. We have $(-1)^{n-1} z_n (z_1 - z_n)(z_2 - z_n) \ldots (z_{n-1} - z_n)$ times the determinant of

$$\begin{bmatrix} z_1 & z_1^2 & \cdots & z_1^{n-1} \\ z_2 & z_2^2 & \cdots & z_2^{n-1} \\ & \cdots\cdots & & \\ z_{n-1} & z_{n-1}^2 & \cdots & z_{n-1}^{n-1} \end{bmatrix}$$

By inductive hypothesis, the latter is

$$z_1 z_2 \dots z_{n-1} \prod_{j < i \leqslant n} (z_i - z_j)$$

Thus the determinant for the $n \times n$ matrix equals the given formula. □

EXAMPLE 6.5.1. The determinant of

$$\begin{bmatrix} a & a^2 & a^3 \\ b & b^2 & b^3 \\ c & c^2 & c^3 \end{bmatrix}$$

is $abc(b-a)(c-a)(b-c)$.

A matrix having this form is called a ***Vandermonde matrix***.

The codes given by the following theorem for $g(x)$ of lowest degree are called ***BCH codes***.

THEOREM 6.5.4. *Let $d, n, q \in \mathbf{Z}^+, d \leqslant n$, and $(n, q) = 1$. Let $\mathbf{F}$ be $GF(q)$. Let $\mathbf{E}$ be a field such that $\mathbf{F} \subset \mathbf{E}$ and there exists $\gamma \in \mathbf{E}$ with $\gamma^n = 1$, but no lower power of γ is 1. Suppose $g(x)$ is a monic polynomial over $\mathbf{F}$ such that $\gamma, \gamma^2, \dots, \gamma^{d-1}$ are roots of $g(x)$, and $g(x) | x^n - 1$. Then the minimum Hamming distance between elements of $g(y)R(y)$ is at least d. Here $R(y)$ denotes the quotient ring where $y = \bar{x}$ and the set $g(y)R(y)$ is the ideal I in $R(y)$ generated by $g(y)$.*

Proof. Let $p(x) \in I$. Then $g(x) | p(x)$ and $p(\gamma) = p(\gamma^2) = \dots = p(\gamma^{d-1}) = 0$. We need to show that if $p(x) \neq 0$ then $p(x)$ has at least d nonzero coefficients. Let

$$p(x) = c_0 + c_1 x + c_2 x^2 + \dots + c_{n-1} x^{n-1}$$

and let $v = (c_0, c_1, \dots, c_{n-1})$. Then the equations $p(\gamma) = p(\gamma^2) = \dots = p(\gamma^{d-1}) = 0$ are equivalent to the matrix equation $vA = 0$ where A is the matrix

$$\begin{bmatrix} 1 & 1 & \dots 1 \\ \gamma & \gamma^2 & \dots \gamma^{d-1} \\ & \dots\dots\dots & \\ \gamma^{n-1} & \gamma^{2(n-1)} & \dots \gamma^{(d-1)(n-1)} \end{bmatrix}$$

We will show that any $(d-1)$ rows of this matrix are linearly independent.

Such rows form a matrix

$$\begin{bmatrix} z_1 & z_1^2 & \dots & z_1^{d-1} \\ z_2 & z_2^2 & \dots & z_2^{d-1} \\ & \dots\dots\dots & & \\ z_{d-1} & z_{d-1}^2 & \dots & z_{d-1}^{d-1} \end{bmatrix}$$

where z_i are distinct members of $\{1, \gamma, \gamma^2, \dots, \gamma^{n-1}\}$.

By Lemma 6.5.3, this determinant is nonzero. Therefore every $(d-1)$ rows of the matrix are linearly independent. Therefore if $vA = 0$, then v has at least d nonzero entries. This means that if v_1, v_2 are encoded vectors then $(v_1 - v_2)A = 0$ so $v_1 - v_2$ has at least d nonzero entries, so $H(v_1, v_2) \geqslant d$. □

We compute the numbers associated with this code as follows. The numbers n, q are as usual. The number of errors correctible is $\frac{d-1}{2}$. The number m is such that there are q^m elements in the ideal. It will be $n - \deg(g(x))$.

The remaining question, is, what choice of $g(x)$ has minimal degree? Let $s(\gamma, \mathbf{F})$ denote the polynomial of least positive degree of which γ is a root. Then $s(\gamma, \mathbf{F})$, $s(\gamma^2, \mathbf{F}), \dots, s(\gamma^{d-1}, \mathbf{F})$ must divide $g(x)$. So the best choice of $g(x)$ is their least common multiple (l.c.m.).

DEFINITION 6.5.1. Let γ be an element of an extension field of $\mathbf{F} = GF(q)$ such that $\gamma^n = 1$ and n is the least such power. Then the *BCH code associated with* γ is the ideal in

$$\frac{\mathbf{F}[x]}{(x^n - 1)\mathbf{F}[x]}$$

generated by

$$g(x) = \text{l.c.m.}\,(s(\gamma, \mathbf{F}), s(\gamma^2, \mathbf{F}), \dots, s(\gamma^{d-1}, \mathbf{F}))$$

where $s(\gamma^i, \mathbf{F})$ is the minimum polynomial of γ^i over $\mathbf{F}$.

EXAMPLE 6.5.2. Let $q = 2, n = 3$. Then $x \in GF(4)\backslash GF(2)$ satisfies $x^3 = 1$. Its minimum polynomial is $x^2 + x + 1$. This is $s(x, \mathbf{F})$. And $s(x^2, \mathbf{F})$ is the same. So

$$\begin{aligned} g(x) &= \text{l.c.m.}\,(x^2 + x + 1, x^2 + x + 1) \\ &= x^2 + x + 1 \end{aligned}$$

The ideal has two code words 0 and $x^2 + x + 1$. This gives the code 000, 111 which can correct 1-error.

If we take $n = q - 1$ we obtain the ***Reed-Solomon codes.*** For these we will have $\gamma \in \mathbf{F}$ is any generator of the multiplicative group, and $s(\gamma^i, \mathbf{F}) = x - \gamma^i$. Therefore

$$g(x) = \prod_{i=1}^{d-1} (x - \gamma^i)$$

Reed-Solomon codes have been used in photodigital memory systems.

A simple encoding function for BCH codes takes $p(x)$ to $p(x)g(x)$. This could be decoded by dividing $g(x)$ into the answer if no errors occurred. However, the standard coding function takes $f(x)$ to $x^{n-m}f(x) - r(x)$ where $r(x)$ is the remainder if $x^{n-m}f(x)$ is divided by $g(x)$.

If errors occur decoding is much more complicated because the error is not a linear function of the received word. However, addition of any word in the ideal does not change the error. E. Berlekamp in 1968 discovered a rapid but somewhat complicated algorthim for correcting errors with BCH codes. See L. Dornhoff and F. E. Hohn (1978).

In many cases, such as equipment failure, errors may not occur at random but as a sequence of adjacent digits. Such an error is called a ***burst-error.*** For given n, m, q the size of a burst-error correctible exceeds the number of random errors correctible. Reed-Solomon codes are good for combinations of random and burst-errors.

A class of codes specially designed for burst-errors are the Fire codes. Let β be the maximum length of a burst which will be corrected.

DEFINITION 6.5.2. Let $c \geqslant \beta$ and let $p(x)$ be an irreducible polynomial of degree c over $GF(q)$. Let s be the smallest positive integer such that $p(x) | x^s - 1$. Assume $(s, 2\beta - 1) = 1$. Then the *Fire code associated with* $p(x)$ is the cyclic code generated by $(x^{2\beta-1} - 1)p(x)$.

For a Fire code, $n = s(2\beta - 1), m = n - 2\beta - c + 1$. It can be verified that $p(x)(x^{2\beta-1} - 1) | (x^s - 1)(x^{2\beta-1} - 1) | x^n - 1$ so we have a suitable generator.

EXAMPLE 6.5.3. Let $q = 2$, $c = 3$, $\beta = 3$. Take as irreducible polynomial $x^3 + x + 1$. The exponent s is 7 since in the field $GF(8)$ generated by a root x, $x^7 = 1$ and no lower power is 1.

The Fire code is generated by $(x^5 - 1)(x^3 + x + 1)$ and n is 35, m is 27. Bursts of length 3 can be corrected.

THEOREM 6.5.5. *The Fire code corrects bursts of length* $\leqslant \beta$.

Proof. Suppose a code word $c(x)$ has a burst-error $b(x)$ during transmission so that $r(x) = c(x) + b(x)$ is received. Write $b(x) = x^j k(x)$ where $k(x) = \Sigma\, k_i x^i$ of degree at most $\beta - 1$ and $k_0 \neq 0$, or $b(x)$ is identically zero.

Then the remainders of $r(x)$ on division by either $x^{2\beta-1}-1$ or $p(x)$ depend only on $b(x)$ since $c(x)$ is a multiple of the generator $(x^{2\beta-1}-1)$. First divide by $x^{2\beta}-1$. This reduces all powers of $b(x)$ modulo $2\beta-1$. So we obtain $x^h k(x)$ where h is the remainder if $2\beta-1$ is divided into j and all powers of k are taken the same way. This is a cyclic rotation of $k(x)$. Now h can be recovered: either it is zero or $x^h k_0$ is the first nonzero coefficient preceded cyclically by $\beta-1$ zero coefficients. Then we recover $k(x)$ by multiplying by x^{-h}.

So the error could only be one of the form $x^{h+(2\beta-1)t}k(x)$ where $t=0,1,2,\ldots,s-1$. It will suffice to verify that any two of these leave different remainders on division by $p(x)$.

Suppose not. Then $p(x)$ divides a difference

$$x^{h+(2\beta-1)t}k(x)-x^{h+(2\beta-1)u}k(x),\ t>u$$

Then

$$p(x)|x^h x^{(2\beta-1)u}(x^{2(\beta-1)(t-u)}-1)k(x)$$

However, $p(x)$ is relatively prime to x since it is irreducible and to $k(x)$ since

$$\deg(k(x))\leqslant\beta-1<\deg(p(x))$$

So

$$p(x)|x^{(2\beta-1)(t-u)}-1$$

But also $p|x^s-1$. And s is relatively prime to $(2\beta-1)(t-u)$ since $0<t-u<s$. This is not possible: if $(2\beta-1)(t-u)=sw+a$ where $a<s$ is the remainder on dividing by s, then

$$x^a\equiv 1(\text{mod } p(x))$$

but s is the least positive integer with this property. □

EXERCISES

In the following exercises, as in the text, q is the number of possible letters or symbols, m the length of a message block before encoding, n the length of words in the code, d the number of errors detectible, t the number of errors correctible. For a Fire code β is the size of a burst-error correctible, and c is the degree of a polynomial $p(x)$ such that $p(x)(x^{2\beta}-1)$ generates the code.

Level 1

1. For $d=n$ what is the only linear code which has Hamming distance $>d$ between any two members?
2. For $d=n-1$ what is the only code with at least two members?
3. For $q=p$ where p is a prime number, let γ generate the extension field $\mathbf{Z}(p^k)$ and be a primitive root. Give an example.
4. For $q=2$, $n=5$, there exists $\gamma\in GF(16)$ satisfying x^5-1. Its minimum polynomial is $x^4+x^3+x^2+x+1$. What code results?

5. For $q = 2$, $n = 7$ there exists $\gamma \in GF(8)$ satisfying $x^7 = 1$ with minimum polynomial $x^3 + x + 1$. Its square satisfies the same minimum polynomial since $(x^3 + x + 1)^2 = x^6 + x^2 + 1$. What is the generator? There exist 8 code words in this code. Write several.
6. Encode 0 0 1 in the code of the above exercise.
7. Decode 0 1 0 1 1 1 0 in this code (no errors occur).
8. Decode 1 0 1 0 1 1 1 (which has errors) by finding a code word within Hamming distance 1.
9. For any linear code (a subspace of $V(n, q)$) explain why the minimum Hamming distance between two members equals the least weight of a member.

Level 2

1. Suppose γ generates the extension field $GF(q^n)$ and is a primitive nth root of unity. For $d = 1$, what is m for BCH codes?
2. Do the BCH codes include any with the same parameters as the Hamming codes?
3. Why must any two of the $s(\gamma^i, \mathbf{F})$ be equal or be relatively prime?
4. Compute a generating polynomial for a Fire code with $\beta = 2$, $q = 2$, $n = 25$, $c = 4$. What is m and the rate?
5. Compute a generating polynomial for a Fire code with $\beta = c = 3$, $q = 2$, $n = 35$. Use the same $p(x)$.
6. Compute a generating polynomial for a Reed–Solomon code correcting one error for $n = 7$. Here γ is any member of $GF(8)\backslash GF(2)$.
7. Encode 1 1 1 0 0 0 in the preceding code.

Level 3

1. Prove, using the idea of Theorem 6.5.4, that for a BCH code and a word $r(x) = c(x) + e(x)$ where $c(x)$ is a code word and $e(x)$ an error having at most t nonzero coefficients the $2t$ quantities $S_i = r(\gamma^i)$, $i = 1, 2, \ldots, 2t$ depend only on $e(x)$ and uniquely determine $e(x)$. This is one basic idea in decoding BCH codes. It reduces the problem of finding the error to solving the $2t$ equations

$$S_i = \sum_{i=1}^{t} c_j \gamma^{in_j}$$

for c_i, n_i.
2. Prove the g.c.d. over any field of $x^i - 1$ and $x^j - 1$ is $x^k - 1$ where k is the g.c.d. of i, j.
3. Explain how to find prime polynomials of low degree over a field $GF(q)$ by a ***sieve method***: find all irreducible polynomials of degrees $< n$. Divide a polynomial of degree p by all irreducible polynomials of degree $\leqslant p$. If $p(x)$ is irreducible, so is $p(ax + b)$, $a \neq 0$.

4. Find all irreducible monic polynomials of degrees 3, 4 over $GF(3)$ of degrees 4, 5 over $GF(2)$.
5. Construct codes using the previous polynomials.
6. Find a code correcting 2-errors, having rate less than 2.

6.6 LATIN SQUARES

Orthogonal Latin squares are used in experiments in which several factors are tested simultaneously, and to counteract the influence of extraneous factors.

Suppose we wish to examine under which conditions a plant will grow best, in a certain region of the country. We want to find the best nutrient, pest-control method, amout of water needed, and amount of light needed. We will test 5 variations on each one. If we are to try all possible combinations of factors, we would need $5^4 = 625$ different experiments. However, with only 25 we can arrange the experiments so that every pair of factors occur in all possible combinations with each other. (In fact we will later show we could even do this with 6 different factors.)

Consider the matrix

$$\begin{bmatrix} (1,1) & (2,2) & (3,3) & (4,4) & (5,5) \\ (2,3) & (3,4) & (4,5) & (5,1) & (1,2) \\ (3,5) & (4,1) & (5,2) & (1,3) & (2,4) \\ (4,2) & (5,3) & (1,4) & (2,5) & (3,1) \\ (5,4) & (1,5) & (2,1) & (3,2) & (4,3) \end{bmatrix}$$

For each pair i, j make an experiment with nutrient i, pest-control method j, and let the (i, j)-entry of this matrix determine the amounts of water and light received by the plants.

The conditions that every pair of factors occurs exactly once means that in each row and in each column every number from 1 to 5 occurs exactly once as a first number and exactly once as a second number, and that every ordered pair (i, j) occurs somewhere in the matrix. These conditions are satisfied in the case above.

DEFINITION 6.6.1. A $n \times n$ ***Latin square*** is a matrix whose entries are numbers from 1 to n such that every number occurs at least once in every row and at least once in every column.

EXAMPLE 6.6.1. Let A be

$$\begin{bmatrix} 1 & 2 & 3 \\ 2 & 3 & 1 \\ 3 & 1 & 2 \end{bmatrix}$$

This is a Latin square.

DEFINITION 6.6.2. Two Latin squares A, B are *orthogonal* if and only if the set of ordered pairs (a_{ij}, b_{ij}) includes every ordered pair of integers from 1 to n.

EXAMPLE 6.6.2. Let A be the matrix of first entries in the 5×5 matrix above and B be the matrix of second entries.

$$\begin{bmatrix} 1 & 2 & 3 & 4 & 5 \\ 2 & 3 & 4 & 5 & 1 \\ 3 & 4 & 5 & 1 & 2 \\ 4 & 5 & 1 & 2 & 3 \\ 5 & 1 & 2 & 3 & 4 \end{bmatrix}, \begin{bmatrix} 1 & 2 & 3 & 4 & 5 \\ 3 & 4 & 5 & 1 & 2 \\ 5 & 1 & 2 & 3 & 4 \\ 2 & 3 & 4 & 5 & 1 \\ 4 & 5 & 1 & 2 & 3 \end{bmatrix}$$

Then A, B are orthogonal Latin squares.

THEOREM 6.6.1. *Let $n = p_1^{n_1} p_2^{n_2} \ldots p_k^{n_k}$. Let $k = \inf \{p_i^{n_i} - 1\}$. Then there exist k mutually orthogonal $n \times n$ Latin squares.*

Proof. We form a ring R which is the direct product of $GF(p_i^{n_i})$. Then R is a commutative ring of order n having a unit $(1, 1, \ldots, 1)$. Choose k nonzero elements x_{ij} from $GF(p_i^{n_i})$ for $i = 1, 2, \ldots, k$, $j = 1, 2, \ldots, k$. Then the elements $y_j = (x_{1j}, x_{2j}, \ldots, x_{nj})$ have the property that y_j is invertible and for $i \neq j$, $y_j - y_i$ is invertible. Write $y_{k+1}, y_{k+2}, \ldots, y_n$ for the remaining elements of R.

Form k $n \times n$ Latin squares $M\langle r\rangle$ by $m\langle r\rangle_{ij} = y_r y_i + y_j$. This is a 1-1 function in i, j separately since y_r has an inverse. Therefore each row and each column ranges over all elements $y_1, y_2, \ldots, y_n$.

Consider $M\langle r\rangle$ and $M\langle s\rangle$. If they are not orthogonal then for some i, j, u, v where $(i, j) \neq (u, v)$ the two matrices have (i, j)-entries equal to their (u, v) entries. Therefore

$$y_r y_i + y_j = y_r y_u + y_v, \quad y_s y_i + y_j = y_s y_u + y_v$$

So

$$(y_r - y_s) y_i = (y_r - y_s) y_u$$

So $y_i = y_u$ since $(y_r - y_s)^{-1}$ exists. So $y_v = y_j$ also. This contradicts $(i, j) \neq (u, v)$. □

COROLLARY 6.6.2. *If n is a prime power, then there exist $n - 1$ mutually orthogonal $n \times n$ Latin squares.*

EXAMPLE 6.6.3. For $n = 5$, we take square 1 having (i, j)-entry $i + j$ modulo 5 and square 2 having (i, j)-entry $i + 2j$ modulo 5. This gives

$$\begin{bmatrix} 0 & 1 & 2 & 3 & 4 \\ 1 & 2 & 3 & 4 & 0 \\ 2 & 3 & 4 & 0 & 1 \\ 3 & 4 & 0 & 1 & 2 \\ 4 & 0 & 1 & 2 & 3 \end{bmatrix}, \begin{bmatrix} 0 & 1 & 2 & 3 & 4 \\ 2 & 3 & 4 & 0 & 1 \\ 4 & 0 & 1 & 2 & 3 \\ 1 & 2 & 3 & 4 & 0 \\ 3 & 4 & 0 & 1 & 2 \end{bmatrix}$$

For prime powers, this result is best possible. For any number n there exists a Latin square, the addition table of $\mathbf{Z_n}$. For $n = 1, 2, 6$; G. Tarry in 1900 proved there does not exist any pair of orthogonal Latin squares. In 1782 Euler had proved already there exist a pair of orthogonal Latin squares for all $n \not\equiv 2 \pmod 4$. The problem of whether orthogonal Latin squares exist for $n \equiv 2 \pmod 4$ was unsolved until the 1958 work of R. C. Bose, S. S. Shrikhande, and E. T. Parker.

Many constructions for Latin squares depend on an equivalent concept, ***orthogonal array***.

DEFINITION 6.6.3. A $k \times m$ matrix A whose entries are taken from a set S is an *orthogonal array* provided that each pair of rows is orthogonal. Two rows A_{i*}, A_{j*} are *orthogonal* if and only if the ordered pairs (a_{is}, a_{js}) are distinct for $s = 1$ to m.

Here we are concerned with the case $|S| = n$, $m = n^2$.

EXAMPLE 6.6.4. Any set of mutually orthogonal Latin squares gives an orthogonal array if we write the entries of each matrix as a single row in the array.

EXAMPLE 6.6.5. For the Latin squares

$$\begin{bmatrix} 0 & 1 & 2 \\ 1 & 2 & 0 \\ 2 & 0 & 1 \end{bmatrix}, \begin{bmatrix} 0 & 1 & 2 \\ 2 & 0 & 1 \\ 1 & 2 & 0 \end{bmatrix}$$

we obtain the array

$$\begin{bmatrix} 0 & 1 & 2 & 1 & 2 & 0 & 2 & 0 & 1 \\ 0 & 1 & 2 & 2 & 0 & 1 & 1 & 2 & 0 \end{bmatrix}$$

by writing out each matrix row by row. The columns include all 9 possible ordered pairs each exactly once.

For two rows to be orthogonal means as here that the corresponding entries of the rows run through all entries of the Cartesian product set.

THEOREM 6.6.3. *There exist m mutually orthogonal Latin squares if and only if there exists an orthogonal array of size* $(m + 2) \times n^2$ *whose entries are from* $\{1, 2, \ldots, n\}$.

Proof. These two matrices are orthogonal

$$A = \begin{bmatrix} 1 & 1 \ldots 1 \\ 2 & 2 \ldots 1 \\ & \ldots\ldots \\ n & n \ldots n \end{bmatrix}, \quad B = \begin{bmatrix} 1 & 2 \ldots n \\ 1 & 2 \ldots n \\ & \ldots\ldots \\ 1 & 2 \ldots n \end{bmatrix}$$

and a matrix is a Latin square if and only if it is orthogonal to both A and B. Therefore from m mutually orthogonal Latin squares $M\langle i \rangle$ we obtain an $(m + 2) \times n^2$ orthogonal array by writing $A, B, M\langle 1 \rangle, M\langle 2 \rangle, \ldots, M\langle m \rangle$ each as a single array.

Conversely suppose we have an $(n + 2) \times n^2$ orthogonal array. Then each row must include every integer $1, 2, \ldots, n$ exactly n times. Rearrange the columns so that row 1 is

$$[1\ 1\ 1 \ldots 1\ 2\ 2 \ldots 2 \ldots n\ n \ldots n]$$

Now row 2 is orthogonal to row 1. So underneath $[i\, i \ldots i]$ the entries of row 2 are $[1\, 2 \ldots n]$ in some order. Rearrange these so they are $[1\, 2 \ldots n]$. Then rows 1, 2 are

$$\begin{bmatrix} 1\ 1\ 1 \ldots 1\ 2 \ldots n\ n \ldots n \\ 1\ 2\ 3 \ldots n\ 1 \ldots 1\ 2 \ldots n \end{bmatrix}$$

That is, they are A, B. So the other rows regarded as matrices, are matrices orthogonal to A, B and to each other. So they are mutually orthogonal Latin squares. □

EXAMPLE 6.6.6. If we add A, B to the previous orthogonal array we have

$$\begin{bmatrix} 0 & 0 & 0 & 1 & 1 & 1 & 2 & 2 & 2 \\ 0 & 1 & 2 & 0 & 1 & 2 & 0 & 1 & 2 \\ 0 & 1 & 2 & 1 & 2 & 0 & 2 & 0 & 1 \\ 0 & 1 & 2 & 2 & 0 & 1 & 1 & 2 & 0 \end{bmatrix}$$

DEFINITION 6.6.4. Two Latin squares are *isotopic* if one can be obtained from the other by changing the labels of elements, permuting the rows, and permuting the columns.

EXAMPLE 6.6.7. The squares

$$\begin{bmatrix} a & b & c \\ b & c & a \\ c & a & b \end{bmatrix}, \begin{bmatrix} c & b & a \\ a & c & b \\ b & a & c \end{bmatrix}$$

are isotopic. Change labels a, b, c to c, b, a. Then interchange the last two rows.

Isotopy is an equivalence relation on Latin squares. The number of isotopy classes of $n \times n$ Latin squares for small values of n is:

n	Isotopy classes
1-3	1
4-5	2
6	22
7	563
8	1,676,257

EXERCISES

Level 1

1. Construct two orthogonal 3×3 Latin squares.
2. Construct four orthogonal 5×5 Latin squares.
3. Show any 3×3 Latin square can be written as

$$\begin{bmatrix} a & b & c \\ b & c & a \\ c & a & b \end{bmatrix}$$

by possibly interchanging two columns.
4. Write out the three mutually orthogonal Latin squares corresponding to $GF(4) = \{0, 1, x, y = 1 + x\}$ where addition is modulo 2 and multiplication satisfies $x^2 = y, x^3 = 1$.
5. Give two 4×4 Latin squares which are not isotopic.
6. Prove that if $n \not\equiv 2 \pmod 4$ and $n \not\equiv 3, 6 \pmod 9$, there exist at least three mutually orthogonal $n \times n$ Latin squares.
7. Given orthogonal arrays A, B of size $k \times n^2$ and $k \times m^2$ from sets of n, m elements, construct a $k \times n^2m^2$ orthogonal array, whose entries are ordered pairs. Each row should be the Cartesian product of the corresponding rows of A, B listed in a fixed order. Give an example for $k = 2, n = 2, m = 2$.

Level 2

1. Show that r rows of a Latin square can be extended to $r+1$ rows if and only if the system formed by the complements of the columns has an SDR.
2. Prove there exist exactly two isotopy classes of 4×4 Latin squares.
3. Give two 5×5 Latin squares which are not isotopic.
4. Let G be any set with a binary operation denoted product, such that for all a, b there exists x, y with $ax = b$ and $by = a$. Show that the multiplication table of G is a Latin square.
5. Count the number of 3×3 Latin squares with entries a, b, c. Count 4×4 Latin squares with entries labelled a, b, c, d.
6. Prove there is a correspondence between sets of t mutually orthogonal $q \times q$ Latin squares and codes which can detect t-errors with word length $n = t + 2$ where $m = 2$. Let $M\langle r\rangle$ be the Latin squares. Let the coding of i, j be the sequence $m\langle 1\rangle_{ij}, m\langle 2\rangle_{ij}, \ldots, m\langle t\rangle_{ij}$.

Level 3

1. Show there exist at most $m-1$ pairwise orthogonal Latin squares, for any m.
2. For two groups G_1, G_2 when are the Latin squares arising from G_1, G_2 isotopic?
3. Show that for n a prime power $\equiv 3\ (4)$ there exist two orthogonal $m \times m$ Latin squares where $m = \dfrac{3n-1}{2}$ (E.T. Parker). Let g generate the multiplicative group of $GF(n)$. The condition on n guarantees that -1 is not a square in $GF(n)$. Let $y_1, y_2, \ldots, y_{\frac{1}{2}(n-1)}$ be symbols not in $GF(n)$. We construct a $4 \times n^2$ array as follows. Take all columns:

$$\begin{bmatrix} y_i & g^{2i}(g+1)+x & g^{2i}+x & x \\ x & y_i & g^{2i}(g+1)+x & g^{2i}+x \\ g^{2i}+x & x & y_i & g^{2i}(g+1)+x \\ g^{2i}(g+1)+x & g^{2i}+x & x & y_i \end{bmatrix}$$

where x varies through $GF(n)$ and i independently varies through $1, 2, \ldots, \frac{1}{2}(n-1)$. Add n columns

$$\begin{bmatrix} x \\ x \\ x \\ x \end{bmatrix}$$

as x runs through $GF(n)$. Add $\frac{(n-1)^2}{4}$ columns whose entries correspond to any pair of orthogonal $\frac{n-1}{2} \times \frac{n-1}{2}$ Latin squares whose entries are $y_1, y_2, \ldots, y_{\frac{n-1}{2}}$. Note that $\frac{n-1}{2}$ is odd. Prove this is an orthogonal array.

4. Construct a 10×10 pair of orthogonal Latin squares using the method of the previous exercise.

6.7 PROJECTIVE PLANES AND BLOCK DESIGNS

In ***projective geometry***, it is convenient to assume that in addition to the ordinary points of the plane there are points at infinity, one for each class of parallel lines. All the points at infinity lie form a set called the ***line at infinity***. With these additions, in the projective plane, any two lines intersect on a unique point, and any two points lie on a unique line.

However, there also exist other structures with the same property such as a projective plane having 7 lines and 7 points, 3 points on each line, and 3 lines through each point.

DEFINITION 6.7.1. A *projective plane* consists of sets P, L whose elements are respectively called points, lines and a binary relation r from P to L called '*point p lies on line m*', such that

PP1. Two distinct points lie on one and only one line.
PP2. Two distinct lines have one and only one point in common.
PP3. There exist four distinct points no three of which lie on any line.

EXAMPLE 6.7.1. Suppose we take L = {great circles on a sphere}, P = {pairs of diametrically opposite points on a sphere}. Then P, L form a projective plane.

The standard example of a projective plane has P being the set of points of the plane together with 1 point at infinity lying on each line through the origin, and L the lines of the plane, together with the line at infinity, whose points are the points at infinity. A line m in the plane contains one and only one point at infinity, the point for the line through the origin parallel to m.

THEOREM 6.7.1. *For any field or division ring* **F** *exists a projective plane whose points are the 1-dimensional subspaces of a 3-dimensional vector space over* **V** *and whose lines are the 2-dimensional subspaces of* **V**.

Proof. The intersection of any two distinct 2-spaces must be nonzero since otherwise their direct sum. of dimension 4 would be contained in a three-dimensional space. So it is a space of dimension 1, since its dimension is less than 2. Any two distinct 1-spaces generated a 2-dimensional subspace. The 1-spaces spanned by $\{(0,0,1),(0,1,0),(1,0,0),(1,1,1)\}$ satisfy the last condition. □

This theorem gives a projective plane having $\frac{p^{3k}-1}{p^k-1}$ points and $\frac{p^{3k}-1}{p^k-1}$ lines, for every prime power p^k. Such planes can be characterized by the fact that the ***theorem of Desargues*** holds in them.

This theorem states that if the following sets of points are collinear: (a) $0, A_1, A_2$, (b) $0, B_1, B_2$, (c) $0, C_1, C_2$ then the three intersections (d) A_1B_1 intersected with A_2B_2, (e) A_1C_1 intersected with A_2C_2, (f) B_1C_1 intersected with B_2C_2, are collinear.

Many other projective planes exist in which the theorem of Desargues is not true. See Hall (1959), Hughes and Piper (1973), and Pickert (1955).

THEOREM 6.7.2. *Suppose that some line in a projective plane contains $n+1$ points. Then every line contains $n+1$ points, every point is on $n+1$ lines, and there are n^2+n+1 points and n^2+n+1 lines.*

Proof. Let P, Q, R, S be 4 points no three of which are collinear. Then no three of the lines PQ, QR, RS, PS can pass through a single point. By symmetry, it suffices to prove this for PQ, QR, RS. If a point X is on all three lines, it would follow that any two lines coincide, so they are all equal. Thus P, Q, R are collinear, a contradiction. This establishes a perfect duality between the points and lines in results about projective planes.

Consider two lines m_1, m_2. If all points belonged to m_1 or m_2, then P, Q, R, S would, say P, Q on m_1 and R, S on m_2. But then the intersection of PQ, RS could not lie on either without contradiction. Let O not be on m_1 or m_2. Then to X on m_1 we have the point $\overline{OX} \cap m_2$ on m_1. This gives a one-to-one correspondence between the points on m_1 and the points on m_2. So every line has $n+1$ points.

There is also a one-to-one correspondence from points on m_1 to lines through O, given by $X \to OX$. So there are $n+1$ lines through O. By duality to the result about points, there are $n+1$ lines through every point.

Now the entire plane consists of O, and the points other than O on the $n+1$ lines through O. This gives n^2+n+1 points. By duality there are n^2+n+1 lines. □

The number n is called the *order* of the projective plane.

Projective planes in general correspond to ternary systems generalizing fields. A ternary operation on a set S is a mapping $S \times S \times S \to S$. We indicate how this is done. First coordinates in a set of n elements are found. Choose any four points X, Y, O, U no three on a line. Here O is to be the origin, OX the x-axis, OY the y-axis, XY the line at infinity and OU the line of points (a, a).

First, set $O = (0, 0)$, $U = (1, 1)$ and assign coordinates (a, a) to the other points of OU in any 1-1 fashion, where $a = 2, 3, \ldots, n-1$. If $P \notin OU \cup XY$, let $P = (a, b)$ where (b, b) is the intersection of XP and OU and (a, a) is the intersection of YP and OU. To a point at infinity on the line joining $(0, 0)$ and $(1, m)$ assign the coordinate (m) (which can be considered a slope of a class of parallel lines). Assign Y the coordinate (∞).

Then a ***ternary operation*** on $0, 1, 2, \ldots, n-1$ is defined by $y = x \cdot m \bigcirc b$ if and only if the line L through (m) and $(0, b)$ contains the point (x, y). Since this is the intersection of L and the line joining (∞) and (x, x), a unique y exists.

This ternary operation has the following properties:

T1. $0 \cdot m \bigcirc c = a \cdot 0 \bigcirc c = c$

T2. $1 \cdot m \bigcirc 0 = m \cdot 1 \bigcirc 0 = m$

T3. The function $a \cdot m \bigcirc x$ is 1–1 in x for all a, m

T4. For any $m_1 \neq m_2$, b_1, b_2 there exists a unique x such that $x \cdot m_1 \bigcirc b_1 = x \cdot m_2 \bigcirc b_2$

T5. If $a_1 \neq a_2$, c_1, c_2 are given then there exists a unique pair m, b such that $a_1 \cdot m \bigcirc b = c_1$ and $a_2 \cdot m \bigcirc b = c_2$.

Conversely it can be proved that any ternary system with these five properties determines a projective plane.

EXAMPLE 6.7.2. For projective planes associated with subspaces of a 3-dimensional vector space, the ternary ring is isomorphic to the field **F** with ternary operation $x \cdot m \bigcirc b = xm + b$.

Many other systems result in various other types of projective planes, various nonassociative division rings, near-fields, alternative division rings.

Every known projective plane has order a prime power, and the most famous unsolved problem in projective plane theory is, ***do there exist projective planes of order not a prime power***? Existence of a projective plane of order n holds if and only if there exist $n-1$ pairwise orthogonal $n \times n$ latin squares.

Projective planes may also be described in terms of the $(0, 1)$-matrix A which is the matrix of the binary relation $p \in L$ for points p and lines L.

PROPOSITION 6.7.3. *An $s \times s$ $(0, 1)$-matrix A defines a projective plane if and only if $A^T A = AA^T = nI + J$ where J is the $n \times n$ matrix all of whose entries are 1, for $s > 3$.*

Proof. $(A^T A)_{ij} = \Sigma\, a_{ki} a_{kj}$ is the number of points k in the intersection of lines k, j. Thus it should be 1 if $i \neq j$, $n+1$ if $i = j$. And $(AA^T)_{ij} = \Sigma\, a_{ik} a_{jk}$ is the number of lines passing through points i, j. So it should be 1 if $i \neq j$, else $n+1$.

Conversely the matrix conditions imply that (PP1) and (PP2) and that every line has the same number of points and every point is on the same number of lines. By examining the special cases satisfying 1, 2 but not 3 we find that none satisfies the hypotheses of the proposition. □

A ***balanced incomplete block design*** has uses similar to those of Latin squares in statistics. It is used in the design of experiments in which it is not convenient to have to test every value of factor a against every value of factor b, for varying values of factor c. That is, a Latin square gives a pattern of experiments.

$$\text{Factor } b \quad \overset{\text{Factor } a}{\begin{bmatrix} 1 & 2 & 3 \\ 2 & 3 & 1 \\ 3 & 1 & 2 \end{bmatrix}}$$

where the value of factor c is in the interior. A more general balanced incomplete block design would be

$$\text{Factor } b \quad \begin{bmatrix} 1 & 2 \\ 2 & 3 \\ 3 & 1 \end{bmatrix}$$

Here perhaps it is feasible to make only two experiments for each value of factor b. Only two factors, b, c are involved.

DEFINITION 6.7.2. A *balanced incomplete block design* (BIBD) of type (b, v, r, k, λ) consists of a family B_i, $i = 1$ to b of subsets of a set V of v elements such that (1) $|B_i| = k < \lambda$ for all i, (2) $|\{i : x \in B_i\}| = r$ for all $x \in V$, (3) $|\{i : x \in B_i \text{ and } y \in B_i\}| = \lambda$ for all $x \neq y$ in V.

EXAMPLE 6.7.3. If all B_i have a single distinct element we have a BIBD $(n, n, 1, 1, 0)$. For instance

$$\begin{bmatrix} 1 \\ 2 \\ 3 \end{bmatrix}$$

EXAMPLE 6.7.4. A projective plane of order n is a BIBD $(n^2 + n + 1, n^2 + n + 1, n + 1, n + 1, 1)$.

The block B_i is generally considered as the ith set of experiments and V is considered as the set of treatments, or varieties being tested.

To a balanced incomplete block design is associated an ***incidence matrix*** A of size $v \times b$ where $a_{ij} = 1$ if variety i occurs in block j, otherwise $a_{ij} = 0$. This matrix has the property

$$AA^{\mathrm{T}} = \lambda J + (r - \lambda)V$$

and its column sums are each k. Conversely any $v \times b$ $(0, 1)$-matrix with these properties defines a BIBD (b, v, r, h, λ).

PROPOSITION 6.7.4. *For a BIBD,* $bk = vr, \lambda(v - 1) = r(k - 1)$.

Proof. The first equation is a counting on the total number of elements of the design, first by blocks then by varieties. Let v_1 be a given member of V. Then the number of elements of the design lying in a block containing v_1, but not containing v_1 there are themsleves, is $\lambda(v - 1)$ if we count it for each of the $v - 1$ other treatments. If we count it by the blocks, then there are r blocks containing v_1, each having $k - 1$ other treatments. □

Balanced incomplete block designs can be constructed from doubly transitive groups.

DEFINITION 6.7.3. A group G acting on a set X is *doubly transitive* if for all $x \neq w, y \neq z$ in X there exists $g \in G$ such that $gw = y, gx = z$.

EXAMPLE 6.7.5. S_n is doubly transitive.

PROPOSITION 6.7.5. *Let G be any doubly transitive group acting on a set V. Let S be a subset of V. Then the family* B_i *of distinct sets* $g(S), g \in G$ *is a BIBD where* $|V| = v, \dfrac{|G|}{|\{g : g(S) = S\}|} = b, |S| = k$.

Proof. (1) holds since $|g(S)| = |S|$. (2) and (3) are unaffected if we take m copies of each block for any m. Take therefore as blocks all sets $g(S)$ in which each distinct block occurs a number of times equal to $|\{g : g(S) = S\}|$. Then (2) follows from transitivity since $i \in g(S)$ if and only if $h(i) \in h(g(s))$. And (3) follows from double transitivity since $i, j \in g(S)$ if and only if $h(i), h(j) \in h(g(S))$. □

EXAMPLE 6.7.6. The group of invertible matrices acts doubly transitively on the set of 1-spaces of any vector space. Therefore if the blocks are taken as all k-spaces of the vector space regarded as sets of 1-spaces lying in them, we have a BIBD.

There are a number of generalizations of BIBD. One type is used in the construction of two 22×22 orthogonal Latin squares.

EXERCISES

In the following exercises, for balanced incomplete block designs, v is the number of elements in the union of all blocks (rows), b is the number of blocks, k is the number of elements per block, r is the number of blocks a given element belongs to, and λ is the number of blocks a given pair of elements belongs to.

Level 1

1. Find a 7×7 circulant $(0, 1)$-matrix having 3 ones in each row and column, which is a projective plane. Assume these main diagonal entries are all 1 as well as the $(1, 2)$-entry.
2. Do the same for a 13×13 matrix with 4 ones per row. This should be the projective plane arising from $GF(3)$.
3. Construct a BIBD $(n, n, n-1, n, n-2)$ for any n.
4. Construct a BIBD $(\binom{n}{k}, n, \binom{n-1}{k-1}, k, \binom{n-2}{k-2})$ for any n, from all subsets of a set of n elements.
5. Explain why $k < v$ in BIBD.

Level 2

1. Show that a Veblen-Wedderburn system defines a projective plane. This is a set having two binary operations denoted addition and multiplication such that (1) addition is an abelian group, (2) multiplication has a unit 1 and multiplication by any nonzero element is 1-1 on either side, (3) $(a+b)m = am + bm$, (4) if $r \neq s$ $xr = xs + t$ has a unique solution x.
2. From the ternary system corresponding to a projective plane explain how to obtain $n-1$ orthogonal $n \times n$ Latin squares. Show how to do the reverse also.
3. What are the parameters of the BIBD of all k-spaces contained in an n-space over $GF(q)$?
4. Prove that in a BIBD $r > \lambda$.
5. If a system satisfies (PP1) and (PP2) for a projective plane and two distinct lines contain respective points (a, b) and (c, d) not equal to their intersection, show (PP3) holds.
6. Consider a system formed by a line L with $n > 2$ points and single point O not on L, together with all lines OX for $X \in L$. Show (PP1), (PP2) but not (PP3) holds.

Level 3

1. Prove the projective planes arising from lines over $GF(q)$ can be represented by a circulant $(0, 1)$-matrix. Represent the 3-space as $GF(q^3)$ and use the fact that its multiplicative group is cyclic.
2. Write a matrix for a projective plane of order 4.
3. Express the general condition on the first row for a circulant matrix to represent a projective plane.

4. Prove that $b \geqslant v$ for a BIBD. Compute $\det(AA^{\mathrm{T}})$ and note that it is zero if $b < v$ since AA^{T} would have rank $\leqslant b < v$.
5. For all values of b, v, k which are at most 5 such that positive integers λ, r can be defined by $bk = vr$, $\lambda(v-1) = r(k-1)$ and the inequalities $k < v$, $r > \lambda$, $b \geqslant v$, try to construct a BIBD.

Open problems

An *open problem* is a question for which no one has yet been able to provide an answer and prove it. Most of the open problems listed here are famous, and have been worked on for many years by many mathematicians. Partial results have been obtained. This course provides you with enough tools to understand the problems, so we encourage some of you to solve these problems.

1. Find the structure and number of D-classes of B_n.
2. Find a method of enumerating groups of order p^n, if p is prime and $n \in \mathbf{Z}^+$.
3. Find a simple proof (<100 pages) that every simple group has even order.
4. Two matrices A, B over $\mathbf{Z}^+$ are said to be ***shift equivalent*** if and only if there exist matrices R, S over $\mathbf{Z}^+$ such that for some $n \in \mathbf{Z}^+$, $RA = BR$, $AS = SB$, $RS = B^n$, and $SR = A^n$. Characterize shift equivalence in the case of eigenvalues of multiplicity > 1.
5. What lattices are lattices of ideals in a regular ring? Of normal subgroups in a group?
6. Classify representations of finite groups over finite fields of order not relatively prime to the order of the group.
7. Is every finite group a Galois group of an extension of $\mathbf{Q}$ over $\mathbf{Q}$?
8. Find codes which approximate Shannon's bound and can be rapidly decoded.
9. Is there a projective plane of order other than a prime power? How many mutually orthogonal $n \times n$ Latin squares can exist?
10. Devise improved computational methods for finding any of the following: (i) eigenvectors, (ii) factors of Boolean matrices, (iii) factors of $\mathbf{Z}$, or polynomials, and (iv) solution of simultaneous polynomial equations.
11. Provide a useful and computable concept of complexity of a scientific problem.

List of special symbols

CHAPTER 1

CHAPTER 2

C: the set of complex numbers, 68
S_n: symmetric group on n letters, 68
$\mathbf{Z_m}$: residue classes modulo m, 69
finite field, 125
finite ring, 163
quotient ring, 174
xH: left coset of H, 72
Hx: right coset of H, 72
ABC: $\{abc: a \in A, b \in B, c \in C\}$, 72
$a|b$: a divides b, 72
G/H: quotient group, 74
$\mathbf{R}^+$: set of positive real numbers, 78
$\frac{G/N}{K/N}$: quotient of two quotient groups, 79
SDR: system of distinct representatives, 84
C_g: conjugation, 93
$\mathbf{Z^E}$: set of even integers, 94
$\mathbf{Z^O}$: set of odd integers, 94
O_x: orbit of x, 95
$\frac{|G: H|}{[G:H]}$: the number of cosets of H, 95
A_n: alternating group on n letters, 97
$A \boxtimes B$: Kronecker product, 105
$\phi(n)$: Euler's function, 111

CHAPTER 3

V: vector space, 119
$\mathbf{F^n}$: $\mathbf{F} \oplus \mathbf{F} \oplus \ldots \oplus \mathbf{F}$, 119
I: index set, 122
$\langle S \rangle$: span of S, 123
$U \oplus V$: internal direct sum, 129
dim W: dimension of W, 129
$a_{ij}^{(m)}$: (i, j)-entry of mth power of A^m, 134
A_{ij}: (i, j)-block of A, 134
A^{-1}: inverse of A, 137
cof (A): cofactor of A, 145
adj (A): adjoint of A, 145
Tr (A): trace of A, 146
$x \cdot y$: inner product, 157
$\mathbf{C^n}$: $\mathbf{C} \oplus \mathbf{C} \oplus \ldots \oplus \mathbf{C}$, 159

CHAPTER 4

$\mathcal{R}$: ring, 163, 164
c.d.: common divisor, 163
$\mathcal{I}$: ideal, 164
$\mathcal{R}/\mathcal{H}$: quotient ring, 164
$\mathbf{Z}^+$: the set of positive integers, 165
$\mathcal{D}$: integral domain, 165
$\mathbf{F}[x]$: ring of polynomial over **F**, 166
$\mathbf{Q}[x]$: ring of polynomial over **Q**, 166
g.c.d.: greatest common divisor, 166
$(a, b) = 1$: a and b are relatively prime, 168
$\mathcal{E}$: Euclidean domain, 168
$\nu(a)$: degree of polynomial, 168
$x \sim y$: similarity (ring), 173
ω: primitive nth root of unity, 178
$\mathcal{R}_{\mathcal{D}}$: division ring, 180

CHAPTER 5

$\mathsf{F}(\mathsf{G})$: group ring over **F**, 184
$\mathcal{M}$: module, 184
$\mathsf{R}(\mathsf{G})$: group ring over **R**, 185
char (**F**): characteristic of a field, 193
$\mathcal{K}$: maximal proper left ideal, 194
$\mathsf{C}(\mathsf{G})$: group ring over **C**, 195
$M_n(C)$: the set of $n \times n$ matrices over **C**, 195
χ: group character, 197
$A \otimes B$: tensor product, 198
$[G: H]$: index of H in the group G, 95, 203
$\lambda^k(\mathcal{M})$: kth exterior power, 204
$s_k(\mathcal{M})$: kth symmetric power, 204
$\mathcal{U}_n$: the set of $n \times n$ unitary matrices, 205
σ_k: kth elementary symmetric function, 206
r_I: standard representation, 206

CHAPTER 6

BCH: Bose-Chaudhuri-Hoquenghen code, 211
$[\mathbf{E} : \mathbf{F}]$: degree of the extension, 211
$[\mathbf{C} : \mathbf{R}]$: degree of the complex numbers over the reals, 211
$\mathbf{F}(S)$: field generated by S over **F**, 212
$\mathbf{E}(x)$: field generated by x over **E**, 212

$\mathbf{F}(x)$: field generated by x over $\mathbf{F}$, 212
Δ: discriminant function, 217
$GF(n)$: finite field of order n, 226
$V(n, q)$: the set of n-dimensional vector over $GF(q)$, 228
$H(u, v)$: Hamming distance, 228
f_E: encoding function, 228, 231
l.c.m.: least common multiple, 236
$s(x, \mathbf{F})$: the minimum polynomial of x over $\mathbf{F}$, 236
β: maximum length of a burst, 237
BIBD: balanced incomplete block design, 249

References

G. Birkhoff, *Lattice Theory,* American Mathematical Society, Providence, R.I., 1967.

A. H. Clifford and G.B. Preston, *The Algebraic Theory of Semigroups,* American Mathematical Society, Providence, R.I., 1964.

J. M. Crowe, *A History of Vector Analysis,* Notre Dame University Press, Notre Dame, 1967.

T. Donellan, *Lattice Theory,* Pergamon Press, New York, 1968.

L. L. Dornhoff and F.E. Hohn, *Applied Modern Algebra,* Macmillan, New York, 1978.

M. Gardner, Mathematical games, *Scientific American,* **234**/4 (1976), 126–130 and **234**/9 (1976), 210–211.

W. Gilbert, *Modern Algebra with Applications,* Wiley, New York, 1976

B. C. Griffith, V. L. Maier and A. J. Miller, Describing communication networks through the use of matrix-based measures, preprint, 1976.

M. Hall, Jr., *The Theory of Groups,* Macmillan, New York, 1959.

D. R. Hughes and F. C. Piper, *Projective Planes,* Springer-Verlag, New York, 1973.

K. H. Kim, *Boolean Matrix Theory and Applications,* Marcel Dekker, New York, 1982.

R. D. Luce, Semiorders and a theory of utility discrimination. *Econometrica,* **24** (1956), 178–191.

G. Pickert, *Projective Ebenen,* Springer, Berlin, 1955.

I. Rabinovitch, The Scott–Suppes theorem on semiorders, *Journal of Mathematical Psychology,* **15** (1977), 209–212.

D. Scott, Measurement structure and linear inequalities, *Journal of Mathematical Psychology,* **1** (1964), 233–247.

D. Scott and P. Suppes, Foundation aspects of theories of measurement, *Journal of Symbolic Logic,* **23** (1958), 113–128.

P. Suppes and R. Zinnes, Basic measurement theory, in R. D. Luce, R. R. Bush and E. Galanter, editors, *Handbook of Mathematical Psychology I,* Wiley, New York, 1963.

H. C. White, *An Anatomy of Kinship*, Prentice-Hall, Englewood Cliffs, N.J., 1963.

H. C. White, S. A. Boorman and R. L. Breiger, Social structure from multiple networks, I. Blockmodels of roles and positions, *American Journal of Sociology*, **81** (1976), 730-790.

Index

solvable group